# Science Matters
Humanities as Complex Systems

# Science Matters
## Humanities as Complex Systems

### Maria Burguete

*Scientific Research Institute Bento da Rocha Cabral, Portugal*

### Lui Lam

*San Jose State University, USA*

### Editors

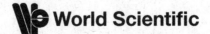

**World Scientific**

NEW JERSEY · LONDON · SINGAPORE · BEIJING · SHANGHAI · HONG KONG · TAIPEI · CHENNAI

*Published by*

World Scientific Publishing Co. Pte. Ltd.

5 Toh Tuck Link, Singapore 596224

*USA office:* 27 Warren Street, Suite 401-402, Hackensack, NJ 07601

*UK office:* 57 Shelton Street, Covent Garden, London WC2H 9HE

**Library of Congress Cataloging-in-Publication Data**
Science Matters : humanities as complex systems / edited by Maria Burguete and Lui Lam.
   p. cm.
  Includes bibliographical references and index.
  ISBN 978-981-283-593-2 (alk. paper)
  1. Science and the humanities. 2. Humanities. 3. Interdisciplinary approach to knowledge.
  4. System theory. 5. Social systems. I. Burguete, Maria. II. Lam, Lui.
  AZ362.S35 2008
  001.3--dc22

                                   2008044676

**British Library Cataloguing-in-Publication Data**
A catalogue record for this book is available from the British Library.

The editors and publisher would like to thank the authors and the publishers of various journals and books for their assistance and permission to reproduce the selected figures found in this volume:
    1. Nature Publishing Group (Figures 12.1 and 12.10)
    2. American Physical Society (Figures 12.2, 12.4, 12.5, 12.6, 12.7, 12.11 and 12.12)

*Cover photograph:* Fishing for knowledge (Lui Lam, Foz do Arelho Beach, 2006)
*Artwork:* Part I: Rounds (Charlene Lam, 2008)
        Part II: Curves (Charlene Lam, 2008)
        Part III: Swoops (Charlene Lam, 2008)

Printed in Singapore.

# Preface

All earnest and honest human quests for knowledge are efforts to understand Nature, which includes both human and nonhuman systems, the objects of study in science. Thus, broadly speaking, all these quests are in the science domain. The methods and tools used may be different; for example, the literary people use mainly their bodily sensors and their brain as the information processor, while natural scientists may use, in addition, measuring instruments and computers. Yet, all these activities could be viewed in a unified perspective: they are scientific developments at varying stages of maturity and have a lot to learn from each other.

That "everything in Nature is part of science" was well recognized by Aristotle and da Vinci and many others. However, it is only recently, with the advent of modern science and experiences gathered in the study of statistical physics, complex systems and other disciplines, that we know how the human-related disciplines can be studied scientifically.

Science Matters (SciMat or scimat) is about all human-dependent knowledge, wherein, humans (the material system of *Homo sapiens*) are studied scientifically from the perspective of complex systems (see Chapter 1). Here, the term "complex systems" means simply "very complicated systems," in the sense adopted by common people. SciMat includes all the topics covered in humanities and social sciences—in particular, art, literature, movie, culture, history, philosophy, science communication and the studies of science.

Traditionally, many of these topics are under the name of "science of x" or "science and x," where x could be culture, art, literature, society, and so on, or even science in the former case. However, x here, from the perspective of SciMat, is already a part of science. These descriptions are thus misleading. For example, by saying "science and culture," it implies that science and culture are two different things, which could be opposing each other. Instead, they are different aspects of the same thing—the effort to understand Nature and a new term "science matters" is called for.

To advance the idea of SciMat, a new discipline, the First International Conference on Science Matters was held in Ericeira, Portugal, May 28-30, 2007, co-chaired by Maria Burguete and Lui Lam. The intention was to bring together experts from art/humanities and sciences, finding out from each other how each person's own discipline is done and exchanging ideas. Hopefully, mutual understanding will be achieved and collaboration across disciplines will result, with the aim to raise the scientific level of the disciplines. This is not easy, but the important first step has been taken.

This book contains contributions from invited speakers of this conference, who are asked to expand their lectures for the general readership of all intellectuals. Two additional articles come from experts who are invited by the editors to contribute, after the conference. The articles, ranging from art to philosophy and history to social science and to physics, are loosely grouped under three parts (see Contents).

We are grateful to the contributors for their professionalism and skills in communicating to the non-experts, and the sponsors of the conference: Centro de Estudos Sociais da Universidade Coimbra, Barclays Bank, Fundação Luso-Americana, Fundação para a Ciencia e Tecnologia, Fundação Oriente, Fundação Calouste Gulbenkian and British Council. Their combined support makes this book possible.

Rio Maior, Portugal        Maria Burguete
San Jose, California        Lui Lam

# Contents Summary

# Contents

---

**PART II   PHILOSOPHY AND HISTORY OF SCIENCE**

**PART III   RAISING SCIENTIFIC LEVEL**

*1*

# Science Matters: A Unified Perspective

*Lui Lam*

What is science? The answer is that "everything in Nature is part of science." On the one hand, what we called "natural science" is actually the science of (mostly) simple systems; they are human-independent knowledge. On the other hand, humanities/social sciences—human-dependent knowledge—belong to the science of complex systems. Demarcation of Nature according to human and nonhuman systems, and the recognition that complex systems are distinct from simple systems allow us to understand the world differently and profitably. For completeness, the nature of simple and complex systems is briefly presented. The origin of the two cultures (made famous by C. P. Snow), humanities and "science," is traced and some confusing issues clarified. While a gap between humanists and "scientists" does exist due to historical reasons, there is no intrinsic gap between humanities/social science and "natural science." If these disciplines look different from each other, it is because they are at various level of development, scientifically speaking. To *properly* bridge the gap and to advance the search for human-dependent knowledge, a new discipline—*Science Matters* (SciMat or scimat)—is introduced. SciMat treats all human-related matters as part of science, wherein, humans are studied scientifically from the perspective of complex systems with the help of experiences gained in physics, neuroscience and other disciplines. Consequently, all the topics covered in humanities and social sciences are included in SciMat. The motivation and concept of SciMat, and a successful example (*histophysics*, the physics of human history) are presented and discussed. Four major implications of SciMat are described. In particular, a new answer to the Needham Question is offered for the first time. This chapter ends with discussion and conclusion.

## 1.1 Introduction

All earnest and honest human quests for knowledge are efforts to understand Nature, which includes both human and nonhuman systems, the objects of study in science. Thus, broadly speaking, all these quests are in the science domain. The methods and tools used may be different; for example, the literary people use mainly their bodily sensors and their brain as the information processor, while natural scientists may use, in addition, measuring instruments and computers. Yet, all these activities could be viewed in a unified perspective: they are scientific developments at varying stages of maturity and have a lot to learn from each other.

That "everything in Nature is part of science" (see Section 1.2) was well recognized by Aristotle and da Vinci and many others. Yet, it is only recently, with the advent of modern science and experiences gathered in the study of statistical physics [Lam, 1998; Paul & Baschnagel, 1999], complex systems [Lam, 1997; 1998] and other disciplines, that we know how the human-related disciplines can be studied scientifically.

Science Matters (SciMat or scimat) is the new discipline that treats all human-related matters as part of science. SciMat is about all human-dependent knowledge, wherein, humans (the material system of *Homo sapiens*) are studied scientifically from the perspective of complex systems. Here, the term "complex systems" means simply "very complicated systems," in the sense adopted by common people; they may not be fractals or chaotic. After all, when fractal and chaotic systems are usually complex, not all complex systems are fractals or chaotic; and there exists no unique and satisfactory definition of complex systems, technically or otherwise [Lam, 1998; 2000]. SciMat includes all the topics covered in humanities and social sciences, with human history as a particular example [Lam, 2008a].

This chapter is organized as follows. The very nature of science is revealed in Section 1.2, followed by an introduction and analysis of the "two cultures" in Section 1.3. Demarcation of everything in Nature according to human and nonhuman systems is introduced in Section 1.4. Section 1.5 discusses simple and complex systems, including a brief introduction of the latter, one of which is the human system. The motivation, concept and an example (histophysics) of SciMat is given in Section 1.6. Four major implications of SciMat, including a new answer to the Needham Question, are presented in Section 1.7. Finally, Section 1.8 concludes this chapter with discussion.

## 1.2 What Is Science?

About 2,600 years ago, Thales (ca. 624 BC-ca. 546 BC) proposed the first "theory of everything": Everything is made of water.[1] (See Fig. 1.1.) Subsequently, Aristotle (384-322 BC) studied various aspects of the universe—astronomy, physics, biology, botany, zoology, logic, ethics, politics, and so on—from the same platform [Llyod, 1970]. In other words, he was interested in almost all the subjects of study existing in universities today. This was not accidental.

Fig. 1.1. Two water-loving philosophers: (a) Thales (ca. 624 BC-ca. 546 BC) from the West, and (b) Guan-tze (?-645 BC) from the East.

---

[1] The Chinese philosopher, Guan-tze (?-645 BC), also favored water [Liu, 2006]. He said, among many water-related utterances, "Human is water. When the 'essences' of male and female combine, water flows and takes shape."

The fragmentation of knowledge into different disciplines is a relatively recent phenomenon, starting only a few hundred years ago. It results more from management convenience than from the intrinsic nature of knowledge itself. *Knowledge knows no separating boundaries.* After all, the highest degree conferred by a university is still called Doctor of Philosophy (not Doctor of Physics, for example), wherein, philosophy means "wisdom"—all kinds of wisdom. As will be explained below, there is a material basis underlying the unified intrinsic nature of knowledge.

Knowledge about our universe or the world could be divided into two groups: those unrelated to humans, and those related to humans. For instance, Newton's three laws of mechanics are *human-independent*. That is, if there were extraterrestrial intelligence (ET), sooner or later, these three laws could also be discovered by them, even though the laws might not be named after Newton. Examples of *human-dependent* knowledge are literature and dance. An ET might not dance like us, because it could have three, not two, legs.

Human-independent knowledge is commonly called "natural science"; human-dependent knowledge, humanities and social science. However, this classification is inaccurate and inappropriate. On the one hand, humans are *Homo sapiens*, a material system consisting of atoms—the same atoms that make up the systems studied in "natural science." Consequently, all human-dependent knowledge is part of natural science, since the objects studied in natural science are *all* material systems.

On the other hand, science is about the study of Nature and a means to understand it in a unified way. Nature consists of everything in the universe—all material systems, humans and nonhumans. The two terms science and natural science are thus identical to each other.[2] It then follows that there could be only one conclusion [Lam, 2006a]:

---

[2] With this understanding, every possible enquiry undertaken would be about Nature. The term "science" in its German sense of *Wissenchaft*—any systematic body of enquiry— and its use in the English language will coincide with each other.

science = natural science
            = physical science + social science + humanities          (1.1)

where "physical science" includes not just physics, but chemistry, biology, and so on.[3] In other words: *Everything in Nature is part of science.* This conclusion was known to the early Greeks. If some of our contemporaries do not know about this, it is because the word science is either misunderstood or misused.

## 1.3  The Origin and Nature of the Two Cultures

Forty-nine years ago, on May 7, 1959, Charles Percy Snow gave the lecture "The Two Cultures and the Scientific Revolution" at Cambridge University [Snow & Collini, 1998].[4] The lecture essentially contains three themes: the distinction and non-communication between the scientific culture and the literary culture in the West, the importance of the science revolution (defined by Snow to mean the application of the

---

[3] In this chapter, "natural science" with quotation marks means the science of mostly inanimate systems, identical to that in conventional usage of these two words; the same goes for "natural scientist."

[4] There are at least two factual errors in this famous article, apparently never pointed out by anyone before. (1) Snow is wrong when he writes, "No, I mean the discovery at Columbia by Yang and Lee" (p. 15 in [Snow & Collini, 1998]). In fact, in the famous paper that earned Lee and Yang the Nobel Prize in 1957, the authors' names appears as Lee and Yang [Lee & Yang, 1957]. (The ordering of the two names in this and other joint papers by the two authors is not a small matter; it plays an important role in the two men's subsequent total breakup of collaboration and friendship [Yang, 1983; Lee, 1986; Jiang, 2002; Zi *et al.*, 2004].) While Lee indeed worked at Columbia University, Yang's address at that time was the Institute for Advanced Study at Princeton, New Jersey (see the address bylines in [Lee & Yang, 1957]). The truth is that Yang has never been associated with Columbia University. (2) A few sentences later, still referring to the work of Yang and Lee, Snow makes another mistake in his sentence, "If there were any serious communication between the two cultures, this experiment would have been talked about at every High Table in Cambridge." In reality, the work of Lee and Yang is purely *theoretical*, which is to point out that there was no experimental evidence supporting or refuting parity conservation in weak interactions at that time. They went on to propose several experiments to settle this issue without predicting the outcome of these experiments. Parity nonconservation was discovered in an *experiment* by Chien-Shiung Wu (1912-1997) [Wu *et al.*, 1957], a colleague of Lee at Columbia University.

"atomic particles," presumably nuclear physics and quantum mechanics), and the urgency for the rich countries to help the poor countries. Very interesting, big themes—but nothing original, as admitted by Snow himself (see "The Two Cultures: A Second Look (1963)" in [Snow & Collini, 1998]).

The lecture generated tremendous interest and much discussion around the world, which helped to earn Snow 20 honorary degrees (mostly from universities outside of England) and carve his name in history. While the other two themes are definitely worth talking about, it is the "two cultures" theme that causes the most controversy and debates. This is not at all surprising. Many in the literary circle felt slighted by Snow in his lecture and had to defend themselves or their profession (see Stefan Collini's "Introduction" in [Snow & Collini, 1998]). And, by definition, literary people are those who can write. Now, the important question, not addressed by Snow himself in his lecture, is this: What is the origin of the two cultures?

### 1.3.1 *Emergence of the Two Cultures*

About ten thousand years ago on earth, the early *Homo sapiens*, our ancestors, started to wonder about the things around them—things in their immediate surroundings and things in the sky. Curiosity serves not just human needs but for those who figure out how things work from their observations, it is a survival skill *via* the evolutionary mechanism according to Charles Darwin (1809-1882).

Among these activities, literature is the description of humans' reflection on and understanding of Nature. Here, Nature includes all (human and nonhuman) material systems, such as falling leaves in autumn, the changing weather and seasons, effect of moonlights on lovers, the way humans treat each other in different spatial and temporal settings, and, quite often, thoughts in one's brain as a function of happenings inside or outside the person's body. When the authors write down all these, they are using their bodily sensors (sighting, touching, smelling, hearing, and so on) as the main detectors and their brain as the major information processor. Apart from that, for latecomers, they could also input information by reading what other writers wrote.

As time went by, the observation and understanding of certain kinds of phenomena progressed faster. The process started with Galileo about 400 years ago. The success results from three crucial and cleaver steps.

1. We pick simple systems (such as a ball rolling down an inclined plane) to study.
2. We do big and daring approximations in constructing theories (for example, approximating the ball by a point particle).
3. We use detectors and information processors other than those from our own bodies.

Consequently, for example, how things fall under the influence of gravity can be predicted and measured with high accuracy. Let us consider the case that the falling object is a human body. The human body falling from a tall building is the same complicated human body described in a piece of literature, but in physics we pretend that it is a point particle (that is, an idealized particle with zero size) in our calculations. This is an approximation; it works because the size of the earth, the source of the gravitational force, is much greater than the size of the human body. Furthermore, we can record the positions of the falling body by digital cameras and compare them with our calculations, with the help of calculators or computers.[5]

This branch of study is now called "natural science," which involves mostly nonliving systems even though living systems (such as humans in free fall and other simpler biological bodies) are not excluded. However, the so-called "natural sciences" are actually "science of simple systems," while all human-dependent studies (humanities/social sciences) are about complex systems since, in fact, a single human being is the most complex system in the universe.

As we just pointed out, "natural science" succeeds because it chooses to deal only with a special subset of phenomena. And literature is stuck with the complicated aspects—such as pride and prejudice—of the complex system called humans.

---

[5] For smaller falling objects, low-tech devices are used to record the positions at regular time intervals. This is routinely done in freshmen physics labs.

As study deepened, specialization became essential and we were left with two distinct groups of practitioners, the writers in the literature profession and what Snow called "scientists" for those working in "natural science." Since writers use their own bodies as tools, only those with supreme bodily sensitivity and suitable hard wiring of neurons in their brains can become good writers, while scientists need other types of quality (such as supreme self-confidence) to succeed. There is no overlap between these two groups of professionals,[6] and we end up with "two cultures"—with a gap in between.[7]

### 1.3.2  *The Gap Today*

The method to close the gap between humanists and "natural scientists" as proposed apparently by Snow is to encourage the literary people to learn some freshmen physics, and the "scientists" to read some good literature.[8] This method is widely adopted in the universities and other places but, in fact, *it is problematic and ineffective.*

To understand this, we have to examine how the gap is formed *presently* in practice (while what described in Section 1.3.1 is how the gap was formed historically), and the very nature of the gap itself.

1.  How the gap is formed today

As correctly pointed out by Snow [Snow & Collini, 1998], the existence of today's gap is due to the design of our education system. Students in high schools and universities are directed too early towards either humanities or "science." In response, the way to bridge the gap is to cancel this early division of students in high schools, and (after the gap is formed) force them to take general-education courses in the universities.

---

[6] This was painfully experienced by Snow himself. In 1932, Snow had to recant publicly his "discovery" of how to produce Vitamin A artificially after his calculation was found faulty. Snow, a trained chemist, decided to leave scientific research completely after this incident and became a novelist (p. xx in [Snow & Collini, 1998]). He indeed made the correct move, judging by later developments in his career.

[7] These days, the two separate groups in the two cultures are commonly understood to be humanists on the one hand and "natural scientists" on the other hand.

[8] As good literature is concerned, unlike the case in science, there is no unique choice suitable for everybody. Reading Shakespeare or Tang poems/Song proses will equally do.

These remedies are actually carried out in some countries and in most universities. However, we are no longer in the early Greek days. Economically there is strong competitive pressure to arm our students early with special professional skills. It is impossible to get all countries, especially the developing countries, to agree on a slow-down schedule in their education systems. And so, to narrow the existing gap, the response is to increase the dose of general-education courses in the universities and the enhancement of popular science activities (reading popular science books in particular) in the society. (See [Lam, 2008b].) But will this work in its present form? And is this effective and necessary?

## 2.  Nature of the gap today

The gap today exists in the form of different knowledge contents picked up by the two camps of people, humanists and "scientists," during their schooling periods and beyond. And this is the rationale behind the proposal to encourage them to read something from the other camp. But (1) this is hard to achieve; (2) it is ineffective even achieved.

To illustrate point (1), let us take the different groups within physics as an example. The fact that physicists can talk to each other is true only to a certain extent. There is not much to talk about between a particle theorist and a condensed matter physicist if the subject is the standard model of particles. But all scientists, be they physicists, biologists or chemists, do share some common knowledge such as the second law of thermodynamics, because this law is a required learning in the college education of these scientists.[9]

Professional activities require high concentration of attention and usually are time consuming, and, especially in the case of science, involve very keen competition. Time is short, for the professionals. Many first-rate scientists do not read books, particularly science books,

---

[9] The second law of thermodynamics is the example used by Snow to test the scientific knowledge of the literary people in a gathering (p. 15 in [Snow & Collini, 1998]). This is in fact quite unfair, because the second law is less universal and useful than people think. It applies only to closed systems and only to their thermodynamic equilibrium states. It applies neither to humans—an open system and the interest of literary people—nor to the expanding "cosmos" as Snow wrongly claimed (p. 74 in [Snow & Collini, 1998]). The reason is that our universe is ever expanding and is never in an equilibrium state [Lam, 2004a]. See [Zhao, 2003] for a detailed discussion.

because what contained in books is usually not fresh enough. Instead, they read research papers that they think might be helpful to their (present or future) work. That is what the scientist had in mind when he, asked by Snow what books he read, replied, "Books? I prefer to use my books as tools." (See p. 3 in [Snow & Collini, 1998].) Tools, here, mean something that will help him to do his research. There is in fact a fair chance that literary books will be read by scientists, for relaxing purpose, for example, when they are in an airplane after attending a conference. But these books are not Shakespeare's. The same goes for the people working in literature or humanities. Why should they read any science book if they cannot find anything there that would help them to do their job? Time is short for them, too.

As for point (2), let us assume for the moment that the humanist now knows something about basic physics and the "scientist" has read some Shakespeare or other great literature, and they meet in a cocktail party. If they ask each other what is new in the other's profession, they will not be able to go too far in their discussion because a sensible opinion in literature or "science" these days requires more knowledge than what is in they possession. Instead, for example, they can converse on Ang Lee's "Brokeback Mountain" (2005) or his other movie, "Crouching Tiger, Hidden Dragon" (2000). These movies' storylines are as deep as Shakespeare's, and perhaps more entertaining.

## 3.  The proper way to bridge the gap

The gap can never be completely closed, nor should it be. What makes the world interesting is diversity; diversity requires some of us to be writers or artists and others physicists, and so on. What we can and should do is to *bridge* the gap. In our final analysis, the non-communication between the two camps is not due to the non-overlap of the people involved, but due to the *absence of any common language or principle in their trades*. Isn't it wonderful to teach every student something, if exist, when they are still in high schools or universities that they could use for the rest of their life no matter what profession they end up with, in humanities or "natural science"? That would guarantee that everybody can communicate with anybody else, in a cocktail party or on the beach, say. Yet, this "something" did not exist in the 1950s when

Snow delivered his lecture;[10] that is why Snow resorted to the *ineffective* remedy in bridging the gap—a remedy that is still being blindly adopted by others presently. Today, fortunately, this "something" does exist. Since the late 1970s, some general principles applicable to almost all disciplines (and thus could serve as the common language mentioned above) have been discovered. Before these general principles can be introduced and appreciated properly, let us look more carefully at what the right-hand side of Eq. (1.1) actually means.

## 1.4 Demarcation According to Human and Nonhuman Systems

The "physical science" listed in Eq. (1.1), historically and as explained in Section 1.3, is mostly about the study of inanimate and simple systems. However, with the advancement of chaos theory[11] and the ubiquity of personal computers, and perhaps also due to the stagnation of research in particle physics (superstring or M theory notwithstanding) [Smolin, 2007], quite a number of physical scientists have turn their attention the other direction, towards systems of larger and larger scales and "discover" complex system (such as those from Cells and up in Fig. 1.2) [Lam, 1998].

Generally speaking, social science consists of anthropology, business management, economics, education, environmental science, geography, government policy, law, psychology, social welfare, sociology, and women's studies.[12]

Philosophy, culture, religion, language, literature, art, music, movie and performing arts make up the humanities, at least most of them. History, by its very nature, could be part of social science, but it is listed in the humanities at some universities such as Stanford University. The aim of literature, music and art in the humanities is to stimulate the human brain—through arrangement of words or colors, sound or speech, or shape of things—to achieve pleasure and beauty, or their opposites,

---

[10] The powerful evolution theory of Charles Darwin does cut across all biological systems, but stops at inanimate systems.

[11] See Section 1.5.2 for an introduction to chaos.

[12] http://www.sosig.ac.uk.

*L. Lam*

*via* the neurons and their connecting patterns [Pinker, 1997].[13] The brains, some sort of computer, of the creator and the receiver at the two ends of this process are heavily involved. Linguistic is the study of the tools involved in written words and speeches, supporting the three disciplines mentioned above.

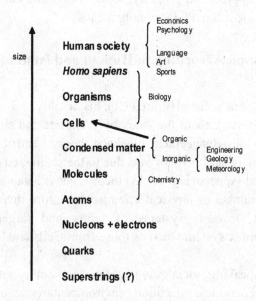

Fig. 1.2. Different systems (left column) studied in various disciplines (right column), listed from bottom to top as the system's size increases.

The *scientific* development of these disciplines in humanities is at a primitive level—the empirical level, using methods that are largely analytic, critical or speculative.[14] And that is why they are separated from social science, which is at an intermediate level (while physical science is at the highest level).

---

[13] And quite often, especially in the case of literature, to promote or stimulate a person's understanding of the world.

[14] http://en.wikipedia.org/wiki/Humanities (July 16, 2008).

At this point, it becomes clear that the three items listed on the right-hand side of Eq. (1.1) are classified distinctly because of their scientific level in development, and *not* because of the nature of the objects under study in each category. Actually, the three items are arranged, from left to right, in increasing level of scientific development.

This classification scheme may be convenient, but is definitely neither logical nor natural. To study something seriously and logically, Nature in this case, one would like to group the objects under study according to their intrinsic nature and not how much we happen to understand them presently. For instance, at least in the beginning, if we want to study orange we will focus our attention on all kinds of oranges and put them in the same category, instead of starting from the category of oranges and clams, say—even though we may compare the two and benefit from the findings of orange and clam studies during the research process. Another example: if we are interested in electrons, we will not group them with rocks and study electrons and rocks together. Common sense, isn't it?

Consequently, to study Nature a natural way is to categorize all objects in Nature into two broad classes, namely, human systems and nonhuman systems.[15] Equation (1.1) then becomes

$$\text{science} = \text{natural science}$$
$$= \text{nonhuman-related science} + \text{human-related science} \quad (1.2)$$

Here, "nonhuman-related science" means the study of inanimate and nonhuman-biological systems—what people usually call "natural science." "Human-related science" consists of humanities, social science and medical science, whereas medical science includes neuroscience and genetics in particular. Apart from the obvious fact that medical science is about humans, the inclusion of medical science in human-related science is dictated by the *belief* and recent findings that many significant human

---

[15] This is due to the fact that it is humans who do the study and control the research budget. If ants were in control, they would classify Nature into the two groups of ants and nonants.

characteristics and behavior (such as morality [Shermer, 2004]) do have a biological basis, as revealed in neuroscience and genetics studies.

By grouping humanities and social science together under one umbrella, human-related science, one can understand anew and more logically the connection between the constituent disciplines (Table 1.1). For the sake of convenience and with full respect for life—an interesting phenomenon in Nature with yet an unknown origin, let us call a human being a "body." There are several basic facts about such a body [Lam, 2002]:

1. Each body is macroscopic, about 40 centimeters to 200 centimeters long; it is a classical particle—that is, quantum mechanics is irrelevant to these bodies.
2. Each body in their daily life moves very slowly compared to light; no need for Einstein's special relativity theory here.
3. The mass of each body is so small (compared to that of a planet, say) that Einstein's general theory of relativity can be forgotten, too.

Table 1.1. Classification of the human system in a focused study according to the number of bodies involved, with examples and major relevant disciplines.

|  | One-body | Few-body | Many-body |
|---|---|---|---|
| *Example* | a Greek male, a Tang Dynasty female, Einstein, Barbra Streisand, Hark Tsui, you, me | Romeo and Juliet, husband and wife, husband and wife living with mother-in-law, a person with two lovers, small-size family, the Beatles | large physics class, tribe, city, country, Roman Empire, society, stock market, IBM |
| *Discipline* | art, music, performing arts, language, literature, psychology, history (biography), neuroscience, genetics, medicinal science, law | psychology, literature, performing arts, history, (family) law | anthropology, (mass) psychology, philosophy, literature, culture, religion, history, business management, economics, education, environmental science, law, social welfare, sociology, women's study, law |

4. Each body consists of layers and layers of structures (molecules, cells, organs, and so on) and many *internal states* (memory, thought, mood, and so on).

5. All bodies are derived from the same ancestor (African Eve, say) only some ten thousand years ago and, according to Charles Darwin's evolution theory, human body and human nature take a long time to evolve and thus are practically unchanged over the last 6,000 years or so—the period in which human history is recorded.

6. Each body is an *open* system, inputting oxygen and food and outputting something else; the second law of thermodynamics does not apply here since the law is for closed systems (and equilibrium states) only [Lam, 2004a].

7. Each body is under the influence of *external fields*, the most important of which is the community or society to which the body happens to belong.[16]

Keeping these facts and Table 1.1 in mind is important and advantageous when a human-related study is being undertaken. It allows you to pick the right tools and the right approximations (that is, simplifying the problem by ignoring some irrelevant factors) to do the research. And it allows you to borrow or be inspired by some successful experience from other areas of study such as physics (wherein, same classification like that in Table 1.1 is used).

For example, in physics, a two-body problem interacting gravitationally with each other[17] is solvable, while such a three-body problem is chaotic and unsolvable [Stewart, 2002]. Now you probably understand why it is so difficult living with your mother-in-law since it is a three-body problem. Just kidding! You cannot simple-mindedly take a

---

[16] Other fields could be physical in nature, such as electromagnetic fields if cell phone is used; sunlight when the body is outdoor (guaranteed in summer in San Jose, California but not necessarily so in Beijing well before *Olympic 2008*); and unavoidably penetrating (harmless) neutrinos and non-penetrating cosmic rays (harmful in large dosage).

[17] The gravitational interaction between two bodies is given by Newton's law of gravity, which states that the force on each body is inversely proportional to the *square* of the separation between the two bodies; similarly, the electric force between two charged bodies due to Coulomb's law.

result from physics and apply it without thinking (or with wrong thinking) to human affairs, because the interacting force between you and your mother-in-law does not obey the inverse-square law as in gravity.[18] More seriously, by ignoring all internal states of a body and treating it like a point particle, and using simple rules of interaction between the particles, computer modeling is able to explain and predict many human-group behaviors, ranging from pedestrian movement and traffic flow to voting processes, economic markets and war [Ball, 2004].

As the classification of systems in Nature is concerned, apart from dividing them into human and nonhuman systems as proposed in this Section, there exists in fact another way, that is, dividing them into simple and complex systems according to the system's complexity. However, as we will show in the next Section, the latter approach though very valuable in clarifying a lot of problems, is not that suitable for the purpose of demarcating systems in Nature.

## 1.5 Simple and Complex Systems

Let us start by explaining what it means to be complex, followed by a brief introduction to complex systems. The reason that complex systems are relevant to our problem of understanding humans will then present itself.

### 1.5.1 *What It Means to be Complex*

There is a feeling that our world is very complex. According to Webster's dictionary:

---

[18] The uncritical application of physical results beyond physics is common, too common, among non-scientists, even among some non-physical scientists. The carefree misuse of chaos in human affairs published in numerous popular science/nonscience books is another example. A further example: Since Einstein's relativity theory tells us that mass and energy can be exchanged, $E = mc^2$; and since my body does have mass, therefore it gives me energy to dance. The fallacy is obvious: a piece of rock also has mass but it does not dance. All these pitfalls lie in the use of analogies without bounds, a symptom of parallelism [Scerri, 1989].

Complex: composed of two or more parts...hard to separate, analyze, or solve.

*Complex* suggests the unavoidable result of a necessary combining and does not imply a fault or failure; *complicated* applies to what offers great difficulty in understanding, solving, or explaining.

Scientifically, the mechanisms driving the complexity of our world are more than what the dictionary suggests; it is not purely due to the existence of two or many parts [Lam, 2004a]. And, in deed, complex systems are hard to analyze, but progress has been made in the last two decades or so [Lam, 1998].

It would be nice if we can quantitatively define *complexity* and use it to compare different situations. Unfortunately, complexity is like love, you know it when you encounter it but it is hard to pin it down. In fact, people came up with more than 30 definitions of complexity, all different from each other. For our purpose, a working definition is enough: *Complexity is determined by the length of the shortest description of a system*; the longer the description, the more complex the system. For example, let us look at three sequences consisting of 1s and 0s.

(a) 111111111111111111111111111111111111
(b) 110010100011000001100101011010100110
(c) 001011100101011100000011010101000111

Sequence (a) can be described by "all 1s." Sequence (b) is generated randomly, and so is described by "1s and 0s generated randomly." There is no pattern in sequence (c),[19] and the description has to be a recitation of the whole sequence; sequence (c) is thus the most complex among the three.

---

[19] Admittedly, this point is hard for the reader to see; to the naked eye, sequences (b) and (c) both look random. However, a technical test can differentiate the two: In the difference map (that is, the plot of $x_n$ vs. $x_{n+1}$ where $\{x_n\}$ is the given sequence of numbers) a random sequence (with large number of elements) will spread out uniformly in the plot, while a non-random sequence will not. Let us say that such a test has been performed.

Such a definition is, of course, problematic. First, if we are not informed that sequence (b) is generated randomly, it could be hard to know this by examining it, and we will assign it the same complexity as that for (c). Second, the definition of the "shortest description" is also subjective. It could be we are not smart enough to identify the rhythm or pattern in (c), or we may be able to identify it a month later. Well, let us keep these limitations in mind and move on.

### 1.5.2 *Complex Systems*

Partly and largely due to the difficulty in defining complexity, there exists no rigorous definition of complex systems. Generally speaking, a complex system usually consists of many interacting components; each component could have a few or many internal states and is adaptive in its behavior. The weakness of such a definition is easy to see. For example, a system may appear complex only because we do not understand it yet. Once understood, it becomes a simple system. Moreover, whether a system is complex or not may depend on what aspect of it we want to study. If we want to know the inner structure and formation mechanism of a piece of rock, the rock could be a complex system. But if we only want to know how the rock will move when given a kick, using Newtonian dynamics will do the job and the rock is simple. The lack of a technical definition and the ambiguity in the concept of simple/complex makes it unsuitable to be used as a demarcation tool of anything.

For our purpose in this chapter a working definition could be adopted: almost all subjects covered in the universities, except those in traditional physics, chemistry and engineering departments, fall into the domain of complex systems [Lam, 1998]. In other words, at the minimum, *biological systems including humans, and all topics covered in humanities and social science belong to the domain of complex systems.* Most of the rest belong to simple systems.[20]

The importance of complex systems makes it worthwhile to know something more about them. Here is some basic knowledge about complex systems.

---

[20] Note that "simple" does not imply "easy"; these are two very different concepts.

All material systems in Nature are made up of "elementary" particles. Going from small to large in size, we have many layers of materials: quarks (or perhaps superstrings), nucleons (protons and neutrons), atoms (nucleons plus electrons), molecules, condensed matter (liquids and solids), cells, human organs, and human beings (Fig. 1.2). It is commonly known that at each layer of organization, there are many *emergent* properties—that is, properties not easily guessed from the lower layer of constituents.

Life is such an emergent property: the fact that a human body is made up of cells and organs does not automatically lend itself to the expectation of life. Another example: the fluidity of water, a property not transparent from knowing that water is made up of $H_2O$ molecules. To describe and understand the emergent property at each layer, one does not need to start from the very bottom level. For example, to describe the flow of water, one does not need to start from the quarks, not even from the molecular level. In fact, based on a few basic principles of symmetry, physicists are able to derive a phenomenological equation describing water flows—the Navier-Stokes equation—which is still being used today. Similarly, to understand complex systems related to human phenomena, one can start from one of several levels such as the *empirical, phenomenological* or *realistic* level [Lam, 2002]. And one has to study complex systems case by case.

Fortunately and surprisingly, since the late 1970s and through the extensive study of simple and complex systems, three general principles of organization in Nature have been discovered. These three unifying principles can be applied to many—though not all—living and nonliving systems,[21] coming, in particular, from humanities and social science. And we are referring to fractals, chaos and active walks [Lam, 1998].[22]

---

[21] Some people call these principles "universal," a misnomer.

[22] Self-organized criticality (SOC) proposed by Per Bak (1948-2002) *et al.* [1987] was advanced as another such general principle for complex systems [Bak, 1996]. Unfortunately, SOC was at odd with many experimental findings in real systems, the ultimate judge in these kinds of things (see, for example, [Cross & Hohenberg, 1993]).

1. Fractals

Fractals were introduced by Benoît Mandelbrot in 1975 [Mandelbrot, 1977]. A fractal is a self-similar (mathematical or real) object, possessing quite often a fractional dimension. Self-similar means that if you take a small part of an object and blow it up in proportion, it will look similar or identical to the original object. A famous example is the Sierpinski gasket [Lam, 2004b]. Fractals are everywhere, ranging from the morphology of tree leaves, rock formations, human blood vessels, to the stock market indices and the structure of galaxies. Fractals are even relevant in the corporate culture [Warnecken, 1993] and the arts [Barrow, 1995].

2. Chaos

Chaos has been investigated by Henri Poincaré at about the turn of the century and subsequently by a number of mathematicians. The modern period occurred in the late 1970s after Mitchell Feigenbaum discovered the "universality" properties of some simple maps, which was preceded by the important but obscure work of Edward Lorentz (1917-2008) [1993] related to weather predictions. Chaos is the phenomenon observed in some nonlinear systems, wherein, the system's behavior depends sensitively on their initial conditions. [23] Examples of chaos include leaking faucets, convective liquids, human heartbeats, and planetary motion in the solar system. The concept is also found applicable in psychology, life sciences and literature [Robertson & Combs, 1995; Hayles, 1991]. A review of chaos for general readers is available [Yorke & Grebogi, 1996].

3. Active walks

Active Walk (AW) is a major principle that Mother Nature uses in self-organization; it is *a generic origin of complexity* in the real world (see also [Zhou *et al.*, 2008]). Active walk is a paradigm introduced by Lui

---

[23] Chaos, a daily word, is used by scientists as a technical word with specific meanings. Before one can call a time sequence of numbers chaotic several tests have to be performed, such as showing the Lyapunov exponent to be positive [Lam, 1998]. The mere look of being random or chaotic is not enough—a pitfall committed by many laypeople.

Lam [2005; 2006a] in 1992 to handle complex systems. In an AW, a particle (the walker) changes a deformable potential—the landscape—as it walks; its next step is influenced by the changed landscape.[24] Active walk has been applied successfully to a number of complex systems coming from the natural and social sciences. Examples include pattern formation in physical, chemical and biological systems such as surface-reaction induced filaments and retinal neurons, formation of fractal surfaces, anomalous ionic transport in glasses, granular matter, population dynamics, bacteria movements and pattern forming, foraging of ants, spontaneous formation of human trails, oil recovery, river formation, city growth, economic systems, parameter-tuning networks [Han *et al.*, 2008] and human history[25] [Lam, 2002; 2006a; 2008a].

These three general principles are what we referred to at end of Section 1.3. All three principles are now an integral part of complex-system science, which is becoming important in the understanding of business, governments and the media, among other things. But, of course, in the study of complex systems there remain a lot of virgin lands waiting to be explored.

## 1.6 Science Matters

The motivation and concept of Science Matters are given here, followed by an example (histophysics). Implications of SciMat will be presented in the next two Sections.

---

[24] For example, ants are living active walkers. When an ant moves, it releases chemicals of a certain type and hence changes the spatial distribution of the chemical concentration. Its next step is moving towards positions of higher chemical concentration. In this case, the chemical distribution is the deformable landscape.

[25] In the AW application in history, think of the walker as an active digger on a soft land. The digger could dig a round trough and keep him moving in circles; he could dig himself a hole deeper and deeper and got himself trapped; or he could dig himself out of a hole and survived. These three situations, respectively, could be used to model what happened to some historical figures; or, when applied sequentially, three stages in the life of an individual. It all depends on the landscaping rule and the stepping rule involved, either or both of which could be time dependent; there are infinite possibilities. That is why AW is such a powerful modeling tool or metaphor in history and other studies.

### 1.6.1  *Motivation*

The discussion presented in Sections 1.2 to 1.5 shows that there is *no* gap between humanities/social science and "natural science"; they are all part of science. After all there exists no natural dividing line among the items listed in Fig. 1.2; it is a continuum. The gap referred to by Snow *is* between humanists and "natural scientists," which was formed historically and is maintained by the education system; this gap is not intrinsic in nature. It is then possible to narrow or bridge *this* gap.

It was almost half-a-century ago that Snow gave his lecture on the two cultures; the world today is quite different. We now realize that the *real* reason to bridge this gap is not simply to let the two sides to have something to converse on in a cocktail party but [Lam, 2006b]:

1.  To have citizens who are better-informed on both humanities/social science and "natural science" and thus can vote more sensibly on issues that could be scientific and/or ethical in nature (such as funding for stem-cell research).

Furthermore, to bridge the gap between humanists and "natural scientists" *properly* and *effectively* as well as, more importantly, to advance *knowledge* about the world and *humanity* (as explained below), we—humanists, social scientists and "natural scientists"—need to work together:

2.  To raise the scientific level of humanities.

While both aims are noble and important, the second one is dearer to us, epistemologically speaking.

### 1.6.2  *Concept*

New disciplines of study are born from time to time, like in the case of human babies, but less frequently; or, like new stars emerging in the sky, being suddenly noticed after a long period in the making.

Science Matters as a new discipline [Lam, 2008c] is created for the two aims listed above. SciMat is the scientific study of all human-related systems. Equation (1.2) is now rewritten as:

science = natural science

= nonhuman-related science + science matters        (1.3)

By naming "human-related science" in Eq. (1.2) as SciMat, we want to emphasize the fact that all human-related matters *are* part of science.

The *concept and method* of SciMat are: Following the good tradition of Aristotle and using the successful experience gained in physics (especially statistical physics), neuroscience and other disciplines, all human-related systems are treated as part of science and studied from the perspective of complex systems. The fact that there do exist general principles (see Section 1.5.2) that cut across all disciplines tells us that this approach is entirely possible.

Figure 1.3 illustrates what we discussed so far. Out of all the objects in Nature, SciMat focuses on the humans (the right box in the upper panel), the most complex system in the universe. Should be fun!

### 1.6.3 *An Example: Histophysics*

History concerns itself with what happened to the *Homo sapiens* in the past [Stanford, 1998].[26] The focus could be on an individual (such as Cleopatra, Alexander the Great or Ava Gardner), a family or an empire; that is, the system under study could be, respectively, one-body, few-body or many-body (Table 1.1). Traditional historians would collect historical records, analyze and put some order in the data or information at hand, then come up with some insights on why something happened and not merely how it happened, and perhaps offer some historical lessons; they stop there usually. No matter how convincing they are, these insights are frequently just educated guesses. As far as I know, no historians in the last few thousands of years had come up with any

---

[26] This is the best book on historiography in my opinion.

historical laws; most of them even doubt the existence of any historical laws.

*Histophysics* [Lam, 2002; 2008a], the physics of human history, is a new discipline that views human history as the past dynamics of a complex system, from the perspective of SciMat; that is, there is a material basis underlying everything happened in human history. History is very complicated or complex, but could be discerned if one is lucky and the right kind of research tools are used. Techniques borrowed from physics and complex systems—such as statistical analysis, computer modeling, computer simulation and the Zipf plot—have been successfully used to tackle problems in history. In particular, *quantitative* laws are found in the distribution of war casualties and of lifetimes of Chinese dynasties (from Qin to Qing, spanning 2,133 years). The latter are in fact laws in macrohistory, favored by the French *Annales* school [Burke, 1990]. (See [Lam, 2008a] for details.)

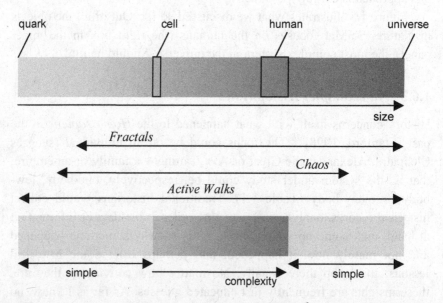

Fig. 1.3. *Upper panel*: systems arranged in increasing size from left to right (not to scale). Humans are the object of study in Science Matters. *Middle three lines*: Range of applicability of fractals, chaos and active walks (from top to bottom). *Lower panel*: Complexity increases from cell to humans, the domain of complex systems; simple systems sit on the left and right regions, respectively, outside of this region.

The success of histophysics confirms the fruitfulness of the SciMat approach in studying human matters; it reinforces our confidence in the direction outlined in SciMat.

## 1.7 Implications of Science Matters

Calling the study of human-related matters by Science Matters is not merely a change in name. There are important implications. Four major implications of SciMat are presented here. More are .given in the next and last Section.

### 1.7.1 *Clearing up Confusion in Terminology*

Let us designate by x the item included in humanities/social science, where x could be art, literature, music, culture, society, and so on. And, for the convenience of discussion, we even allow x to represent "science" in its conventional usage, which actually means "natural science," the science of simple systems. As we all know, there already exist studies of the scientific aspect of x, called the *"science of x,"* say; for example, the "science of art." What is wrong with that?

Nothing, except that it is *confusing*. According to SciMat, art is already part of science. As we explained above, the present state of Art as we know it does not look like a science simply because art, a very complex thing by itself, is at the early stage of its scientific development. Our brain is still the best computer in handling very complex things; that is why artists are still using their brain and not a supercomputer in creating art. Some day in the future, perhaps, when a super-supercomputer better than our brain is available, we will see artists using it to make a living—like the way that physicists are using their personal computer to solve a nonlinear equation these days.

Will it be depressing when this happens to art? Don't worry; the super-supercomputer still needs someone to input something and analyze the results, like the physicist has to decide what nonlinear equation to study, and how to interpret and use the computer results. Anyway, like in the case of physics, one would have more time to go fishing or go to an art museum and be happy.[27]

Why art is developing so slowly as its scientific level is concerned? There are two reasons: (1) Art is a very complex thing and hence it is very difficult to raise its scientific level—the reason just mentioned. (2) And this is very important as art (and in fact arts which include performing arts) goes: humans have buying power; there are enough number of them willing to buy from the artists and thus helping to keep their discipline at a low scientific level. In other words, low-scientific-level art products sell so well already, there is no need to raise its scientific level. It is the market force at work here.

It is interesting to compare this to the case in physics. Physicists are able to control an electron and make it "dance"—a performing (electron) art. The electrons do *not* have buying power; no electron would come and buy a ticket to watch the show.[28] Consequently, the physicists have to raise the scientific level in their trade by doing two things: (1) They build a "superhighway" within a computer chip, which involves Nobel-prize-level breakthroughs. (2) They force the electrons to run like slaves in the superhighway inside the chip. Then they sell this computer chip, electrons included, to their fellow humans. And humans *do* have buying power; you see.

To avoid confusion, the correct way to say "science of art" is "art as a science matter." The case of *"science and x"* is worse; it is *misleading*. For instance, let x be Culture here. "Science and Culture" is misleading because it implies that science and culture are two different things, which could even be opposing each other while, in fact, culture is part of science; culture is a many-body problem in SciMat (see Table 1.1). The use of the term "science and culture" is unfortunate because it endorses Snow's ineffective remedy of bridging the gap (see Section 1.3.2) and thus *prolongs* the gap between the humanists and "scientists." To

---

[27] It seems that people get happy by watching something more complex than what they are doing daily in their lab/office. You never see a physicist leaving her/his lab and go watching the swing of a pendulum. Galileo's days are long gone.

[28] The physicist would be an ultra-billionaire if only 0.001% of the electron population within a 1-mm long copper wire showed up for the show by buying tickets at 0.001 dollars each.

properly bridge the gap, we should help both camps (and everybody else) to understand "culture as a science matter."

## 1.7.2  *The Science Matters Standard*

Since everything is part of science—the fundamental basis of SciMat, there should be one and only one standard in validating the "correctness" of any theory in humanities and social science, that is, the one adopted in "science," established through thousands of years of painful trial and error and the sacrifice of numerous human lives. According to the American Physical Society (*APS News*, June 1999):

> The success and credibility of science anchored in the willingness of scientists to: (1) Expose their ideas and results to independent testing and replication by other scientists. This requires the complete and open exchange of data, procedures and materials. (2) Abandon or modify accepted conclusions when confronted with more complete and reliable experimental evidence. Adherence to these principles provides a mechanism for self-correction that is the foundation of the credibility of science.

With these words in mind, let us propose the following *SciMat Standard*:

1. We will be honest with the reader and ourselves and present our findings in clear writings, and will not try to hide our relevant thinking.
2. We will not quote anyone's writing to support our own argument.
3. We will not be ashamed to admit our own mistakes in our findings and correct them as soon as possible.
4. A conjecture[29] or *hypothesis* becomes a (temporary) *theory* only after it is confirmed by experiments or by practices in the real world.
5. We will abandon (or revise) the theory if it does not agree with confirmed and irrefutable evidences.

---

[29] A mathematical conjecture is a small theorem that is proved. We do not mean this kind of conjecture; the word conjecture used here simply means an educated guess.

Explanations for these five rules are in order. Rule 1 is the basic ethic of any honest researcher or knowledge seeker, but is not always practiced in non-physical disciplines. Yes, we understand that complex systems such as those studied in humanities/social science are very complicated, and one does not always have a clear idea of what one's thought really is. If that is the case, please tell it to your reader which part is clear to you, which part is not, and mark your work as "work in progress." Better still, present your ideas like these in a seminar or cocktail party but not in a conference. If every paper was written clearly and findings/results were presented as "objectively" as possible, and if the paper *always* ended with a section of Discussion/Conclusion in which the author presented what lessons she/he had learned, perhaps the Sokal hoax [Sokal & Bricmont, 1998] would never have to happen and the Science Wars [Labinger & Collins, 2001] could be avoided.

Quoting others to support one's argument is a common practice in non-physical disciplines. But this is completely useless. For example, while Einstein was proven right in his many writings such as the two theories of relativity, he could not always be right and he did not [Kennefick, 2005]. We all know about this; that is why Rule 2.

One should not be ashamed of making mistakes when doing complex-system studies since the job itself is so difficult. What one should do is simply admit their mistakes once recognized and correct them as soon as possible [Shermer, 2001]. In this regard, good economists are real scientists who know their limits and act accordingly; they keep on adjusting their predictions of the stock index or the gross domestic product (GDP) and should be respected for doing that. That explains our Rule 3.

If anyone put out an educated guess (what we mean by hypothesis), this guess has to be confirmed before it can be called a theory. Common sense, right?! Rule 4 is copied from the practice in physics and other "natural sciences." We just want to unify our terminology in communicating to each other, since in SciMat we very likely are coming from different disciplines with different training and background.

Let me emphasize this: We do not mean that physical science is superior to humanities/social science. It is not. In fact, the opposite could be true.[30] Humanists and social scientists are tackling very complex systems, while most physical scientists are dealing with simple systems. Those dealing with complex problems could be more courageous and should be respected. In fact, to be a *good* artist is more difficult than being a good physicist. In physics, there are rules to be obeyed and experience to follow and the choices in solving any physical problem are more restricted[31] than what is available to a painter who wants to create something new. The painter has infinite possibilities and really needs imagination and talent. That is why there are more good physicists than good painters in the world.

With Rule 4 in place, no social theory in the form of political ideology of *any* kind could be validated, since it is unethical to try experiments on living human beings, especially in large numbers.[32] Political leaders are advised to try their "experiments" with computer simulations and be prepared to adjust their policies frequent enough.

Rule 5 is obvious. Finally, it follows from the spirit of this SciMat standard that we will adopt a better standard if that becomes available.

### 1.7.3 *There Is Always the Reality Check*

There is something called the "reality check" as science matters are concerned. We accept Einstein's result, $E = mc^2$, not because it comes from Einstein but because it comes from the special theory of relativity which agrees with most existing experimental findings. Furthermore, the relativity theory gives a sequence of predictions that are later confirmed, even just last month; it helps our confidence in it. The fact that the atomic bomb, built according to $E = mc^2$, works is another plus. This is

---

[30] My daughter is an artist and my hero.

[31] Examples of the restrictions are: all the established laws have to be obeyed and confirmed experimental results respected. New theory cannot ignore or negate them; new theory could improve on them and/or find out where the validity boundary of the old theory is.

[32] Of course, sadly, this has not prevented some historical figures from trying, with disastrous results. An example is what happened in Cambodia: From 1975 to 1979, two million Cambodians were killed under the Pol Pot regime because the leaders mistook a social hypothesis as a theory and applied it to their people; that is, they broke Rule 4 of the SciMat Standard. This and other examples point to the urgent need of greatly improving the scientific training of political leaders, President George W. Bush included.

an example of reality check, even though reality check does not call for an atomic bomb every time, luckily for humans.

Now comes this old woman from Africa who tells us $E = mc$; the reason is that she does not like superscripts. And comes this philosopher from Europe who advocates $E = mc^{\sqrt{2}}$. His reason is that he wants to show his independence from Pythagoras (born between 580 and 572 BC, died between 500 and 490 BC) who abhorred irrational numbers [Lloyd, 1970] and besides, the philosopher honestly thinks the superscript $\sqrt{2}$ is more aesthetically appealing than 2. This phenomenon is called the *multicultural view of science* [Liu, B., 2008]. Who do you think a university will hire into her faculty?

We agree that each one of these three individuals should be fully respected by others for their honest attempts to understand Nature, and be allowed to air their views (freedom of speech) or even publish their findings in a suitable journal/magazine/newspaper of some kind. These days, no opinion can be completely suppressed. At the minimum, there is always the Blog on the Web that they can air their results, and it is free.

In practice, whether someone like you or me would spend our valuable time and listen to an individual's opinion (or read her/his article) on something, science included, does not depend entirely on the quality of that opinion, which we do not know ahead of time anyway. It depends on the *reputation* of that individual; in other words, it is a history dependent process. For instance, we are more inclined to read Einstein's article than that coming from the African woman or the European philosopher, because Einstein has an established reputation. It does not mean, however, everything uttered by Einstein is correct; but even wrong, his uttering could be inspiring. And that is why we make that choice; it is a matter of betting one's time. And we could miss something very valuable and important, because that philosopher's writing could turn out to be very exciting and useful. We accept the risk, and do a catch up by Googling it after the philosopher's finding is reported in the newspaper, say. It is all a matter of allocating finite resources; it has nothing to do with disrespect for Africa or Europe, or *local knowledge* for that matter. The same goes for research funding. Misjudgments are made from time to time; the remedy is to open up more avenues of funding, like those coming from private wealthy individuals or private foundations, just in case.[33]

In fact, people *can* always ignore the reality check and hold on to whatever view of science they want and be happy—and we respect their right to be happy—as long as they do *not* try to put their "theory" to work. To build a cell phone you need quantum mechanics, not any kind of mechanics. But not everyone needs a cell phone, right? And the right of not wanting or producing cell phones should be respected. It is called *cultural diversity*. (See also [Liu, D., 2008].)

### 1.7.4 *The Needham Question*

In 1954, in his book *Science and Civilisation in China* Joseph Needham (1900-1995) [1954] at Cambridge, UK, asked a question that goes something like this: Why did modern science develop only in the West after the 16th century (and not in China who, in the past, applied natural knowledge to practical technology and invention more efficiently than the West did)? [Liu, D., 2008]. There are many explanations offered [Liu, 2000]. Some are obvious; others, not. Here is a new explanation which, I think, is right on the mark.

Remember Aristotle? Aristotle studied and pioneered a number of disciplines, in increasing order of complexity: physics, astronomy, biology and zoology, logic, ethics, government, politics, and so on. Today, his work on physics and astronomy are known to be completely wrong; his biology/zoology is partially wrong; but his logic and ethics studies are still found useful. The same smart Aristotle; how did this happen? The answer is that physics and astronomy are simple systems, biology/zoology is about complex systems, and the rest related to humans are extremely complex systems. It just shows that human-related matters as complex systems are very difficult to study, and we have not made much progress since Aristotle's time in these complex areas. Not Aristotle's fault; we still respect and admire him despite his failures.

---

[33] An example is the funding of extraterrestrial-intelligence (ET) searching in USA. The government funded it for a year and stopped; a rich man came along and continued the funding of SETI, the ET searching institute in Mountain View, CA. Everybody is happy; ET not found up to this moment.

The ancient Chinese—Confucius (551 BC-479 BC) included, for whatever reason (which could be incidental), decided to start their enquiry of Nature with complex systems—humans. They came up with some great insights but no clear conclusions [Wolpert, 1993]. Worse, for unknown reason and unlike Aristotle, they did not or chose not to write down their findings in unambiguous language (that is, they disobeyed Rule 1 in the SciMat Standard; see Section 1.7.2). That is like publishing a paper in a physics journal with writings that the reader can interpret in multiple ways; no way to make progress. In contrast, the Greeks did concern themselves with both simple and complex systems right from the beginning; for example, Archimedes (ca. 287 BC-ca. 212 BC) studied buoyancy of simple bodies and his own body, and was rewarded with the Archimedes Principle. Eventually, the Greek's successful results of simple systems got passed on in the West and ended up in the hands of Galileo, who started modern science. The ambiguous findings in complex systems from ancient Chinese passed on and kept *confusing* and *entertaining* the Chinese for more than two thousand years, even today. This is my answer to the Needham Question.

Here is the Lam Question: Why did modern science arise in Italy and not in other European countries?

## 1.8  Discussion and Conclusion

Here are ten points of interest.

1. There are always grey areas as demarcation of any kind is concerned. Some mathematicians find out about this and come up with a new mathematics called *fuzzy logic* [Klir & Yuan, 1996]. Similarly, the division between humans and nonhumans, and that between simple and complex are not sharp divisions. For example, how many cells have to develop in an embryo before you will call her a human being? In the grey areas, it is common that new and interesting phenomena and question might pop out. Pay attention to the grey areas.

2. According to SciMat, Chinese medicine [Ma, 2007] *is* science at the empirical level. We hope this will settle the debate on this topic once and for all (see [Liu, B., 2008]). The traditional "theory" of Chinese medicine offered in the old books or by its practitioners may seem

strange to outsiders, but they could be some kind of phenomenological theory (or such a "theory" in the making, continuing for over two thousand years)—the next step beyond the empirical level in any scientific development—that works, partially or completely. The fact that the "theory" so far does not match anything in Western medicine implies one of three things: (1) the "theory" is on the wrong track and should be modified or abandoned in the future; (2) the "theory" is on the right track except that it will take time to connect it to that in Western medicine; (3) Western medicine is wrong or irrelevant. Case 3 is unlikely. Case 2 actually happened in the history of superconductivity: the Landau-Ginzburg phenomenological theory (1950) turned out to be correct and could be derived from the BCS microscopic theory (1957) after the latter was discovered [Tinkham, 2004]. Whether that is also the case with Chinese medicine remains to be seen. But we all know this: the debate on Chinese medicine involves something more than prestige. In China, Chinese medicine is heavily funded (quite a number of hospitals and research institutes in Chinese medicine are in place), but anything identified as pseudoscience would be banished.

3. No artists, writers or other humanists should feel threatened by SciMat. They could go on doing what they do best. Humanities are such a vast field that we need a lot of people working on it at the empirical level. Advancing the scientific level of humanities needs to be done mostly by trained "scientists" with the help of or in collaboration with the humanists. We do hope that artists and writers will help.

4. There is no need to abandon the general-education courses in universities. But in all courses, general-education courses in particular, the instructor could start by introducing the concept of SciMat, explaining to the students why Eq. (1.1) should be replaced by Eq. (1.3), and so on. That would help tremendously in narrowing the cultural gap for our students, possibly our future writers and "scientists." Naturally, for the benefit of everybody, we would like to see a SciMat course like *The Real World* [Lam, 2008b] be included in all universities as a *required* general-education course.

5. There could be a brand new theory about macroscopic humans waiting to be discovered, like quantum mechanics lurking there in the case of microscopic systems about 100 years ago. The *only* way to find

out is to set out to look for it, *assuming* temporarily that the new theory indeed exists and is just hidden somewhere, like children searching for Easter eggs on the lawn of the White House each year. What is needed is the smoking gun, similar to Planck's black-body radiation experiment or the double-slit experiment in the case of quantum mechanics.

6. Airplanes and humans, both complex systems, could be very different.

7. When the rule(s) of the SciMat Standard was broken, it often happened that it was humanity and not merely personal honor that suffered.

8. There is already a crowd out there doing econophysics and sociophysics [Chakrabarti *et al.*, 2006; Ball, 2006]; doing humanities as complex systems could be more challenging and rewarding, more fun guaranteed.

9. Da Vinci (1452-1519) could be the last person in history who succeeded in mastering quite a number of topics from both "science"/technology and the arts. His failure to build many of his own designs in engineering, not to mention bringing them to the market, is due to insufficient funding and the absence of a large enough team, and also the non-existence of a suitable industry in society at his time. With the explosion of knowledge in modern times, no one could be as broad and deep as da Vinci was any more. And there is no need to be. What we have to do is encourage people to be experts in two disciplines. With enough number of these bi-disciplinary scholars, all disciplines in the world will be able to link up with each other, directly or indirectly (Fig. 1.4). Here, we are talking about the flourishing of interdisciplinary education and scholarship, and the proper use of science communication [Lam, 2008b], not just for histophysics and SciMat but for all interdisciplinary studies.[34]

10. It was for these reasons that an international conference on SciMat was held in Ericeira, Portugal, May 28-30, 2007 (Fig. 1.5) [Sanitt, 2007]. We are looking forward to more conferences like this one, to

---

[34] In China, there is the journal *China Interdisciplinary Science* (Science Press, Beijing) which treats interdisciplinary studies seriously. It started in 2006 and has published two volumes so far.

provide an international platform for people to exchange ideas face to face. And, learning from the French: to drink, to eat and to sleep [Glover, 2000]. Naturally, the purpose of doing all these is to reach the goal of "Let the Earth be peaceful forever!"

To conclude, Science Matters matter because science matters. But ultimately, Science Matters matter because humans matter!

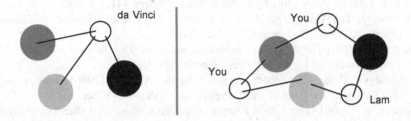

Fig. 1.4. One needs not do everything like da Vinci did (left), but does interdisciplinary work (right). The filled circles represent different disciplines; open circles, individuals.

Fig. 1.5. Poster of the First International Conference on Science Matters, Ericeira, Portugal, May 28-30, 2007.

# References

Bak, P., Tang, C. & Wiesenfeld, K. [1987] "Self-organized criticality: An explanation of 1/*f* noise," *Phys. Rev. Lett.* **59**, 381-384.

Bak, P. [1996] *How Nature Works: The Science of Self-Organized Criticality* (Copernicus, New York).

Ball, P. [2004] *Critical Mass: How One Thing Leads to Another* (Farrar, Straus and Giroux, New York).

Ball, P. [2006] "Econophysics: Culture crash," *Nature* **441**, 686-688.

Barrow, J. D. [1995] *The Artful Universe: The Cosmic Source of Human Creativity* (Little, Brown and Co., New York).

Burke, P. [1990] *The French Historical Revolution: The Annales School 1929-1989* (Cambridge University Press, Cambridge, UK).

Chakrabarti, B. K., Chakraborti, A. & Chatterjee, A. (eds.) [2006] *Econophysics and Sociophysics: Trends and Perspectives* (Wiley-VCH, Berlin).

Glover, W. [2000] *Cave Life in France: Eat, Drink, Sleep...* (Writer's Showcase, Lincoln, NE).

Cross, M. C. & Hohenberg, P. [1993] "Pattern formation outside of equilibrium," *Rev. Mod. Phys.* **65**, 851-1112.

Han, X.-P., Hu, C.-D., Liu, Z.-M. & Wang, B.-H. [2008] "Parameter-tuning networks: Experiments and active-walk model," *Euro. Phys. Lett.* **83**, 28003.

Hayles, N. K. [1991] *Order and Chaos: Complex Dynamics in Literature and Science* (University of Chicago Press, Chicago).

Jiang, C. J. (Chiang, Tsai-chien) [2002] *Biography of Chen Ning Yang: The Beauty of Gauge and Symmetry* (Bookzone, Taibei).

Kennefick, D. [2005] "Einstein versus the *Physical Review*," *Phys. Today*, Sept., 43-48.

Klir, G. J. & Yuan, B. (eds.) [1996] *Fuzzy Sets, Fuzzy Logic, and Fuzzy Systems: Selected Papers by Lotfi A. Zadeh* (World Scientific, Singapore).

Labinger, J. A. & Collins, H. [2001] *The One Culture? A Conversation about Science* (University of Chicago Press, Chicago).

Lam, L (ed.) [1997] *Introduction to Nonlinear Physics* (Springer, New York).

Lam, L. [1998] *Nonlinear Physics for Beginners: Fractals, Chaos, Solitons, Pattern Formation, Cellular Automata and Complex Systems* (World Scientific, Singapore).

Lam, L. [2000] "How Nature self-organizes: Active walk in complex systems," *Skeptic* **8**(3), 71-77.

Lam, L. [2002] "Histophysics: A new discipline," *Mod. Phys. Lett. B* **16**, 1163-1176.

Lam, L. [2004a] *This Pale Blue Dot: Science, History, God* (Tamkang University Press, Tamsui).

Lam, L. [2004b] "A science-and-art interstellar message: The self-similar Sierpinski gasket," *Leonardo* **37**(1), 37-38.

Lam, L. [2005] "Active walks: The first twelve years (Part I)," *Int. J. Bifurcation and Chaos* **15**, 2317-2348.

Lam, L. [2006a] "Active walks: The first twelve years (Part II)," *Int. J. Bifurcation and Chaos* **16**, 239-268.

Lam, L. [2006b] "The two cultures and The Real World," presented at *The 9th International Conference on Public Communication of Science and Technology*, Seoul, May 17-19, 2006, *The Pantaneto Forum*, Issue 24.

Lam, L. [2008a] "Human history: A science matter," in *Science Matters: Humanities as Complex Systems*, eds. Burguete, M. & Lam, L. (World Scientific, Singapore).

Lam, L. [2008b] "SciComm, PopSci and The Real World," in *Science Matters: Humanities as Complex Systems*, eds. Burguete, M. & Lam, L. (World Scientific, Singapore).

Lam, L. [2008c] "Science Matters: The newest and biggest interdicipline," in *China Interdisciplinary Science*, Vol. 2, ed. Liu, Z.-L. (Science Press, Beijing).

Lee, T. D. & Yang, C. N. [1957] "Question of parity conservation in weak interactions," *Phys. Rev.* **104**, 254-258.

Lee, T. D. [1986] *Selected Papers*, Vol. 3 (Birhauser, Boston).

Liu, B. [2008] "Philosophy of science and Chinese sciences: The multicultural view of science and a unified ontological perspective," in *Science Matters: Humanities as Complex Systems*, eds. Burguete, M. & Lam, L. (World Scientific, Singapore).

Liu, D. [2000] "A new survey of the Needham Question," *Studies in the History of Natural Sciences* **18**(4), 293-305.

Liu, D. [2008] "History of science in globalizing time," in *Science Matters: Humanities as Complex Systems*, eds. Burguete, M. & Lam, L. (World Scientific, Singapore).

Liu, D.-C. [2006] *Philosophy of Science* (Renmin University of China Press, Beijing) p. 272.

Lloyd, G.E.R. [1970] *Early Greek Science: Thales to Aristotle* (Norton, New York).

Lorentz, E. [1993] *The Essence of Chaos* (University of Washington Press, Seattle).

Ma, X.-T. [2007] "Understanding Chinese medicine in contemporary cultural context," in *Science Defeated by Superstition?* eds. Jiang, X.-Y. & Liu, B. (Huadong Normal University Press, Shanghai).

Mandelbrot, B. [1977] *Fractals: Form, Chance and Dimension* (Freeman, New York).

Needham, J. [1954] *Science and Civilisation in China*, Vol. 1 (Cambridge University Press, Cambridge, UK) pp. 3-4.

Paul, W. & Baschnagel, J. [1999] *Stochastic Processes: From Physics to Finance* (Springer, New York).

Pinker, S. [1997] *How the Mind Works* (Norton, New York).

Robertson, R. & Combs, A. (eds.) [1995] *Chaos Theory in Psychology and the Life Sciences* (Lawrence Erlbaum, Mahwah, NJ).

Sanitt, N. [2007] "The First International Conference on SCIENCE MATTERS: A unified perspective, May 28-30, 2007, Ericeira, Portugal," *The Pantaneto Forum*, Issue 28.

Scerri, E. R. [1989] "Eastern mysticism and the alleged parallels with physics," *Am. J. Phys.* **57**, 687-692.

Shermer, M. [2001] "I was wrong," *Sci. Am.*, Oct.

Shermer, M. [2004] *The Science of Good and Evil: Why People Cheat, Gossip, Care, Share, and Follow the Golden Rule* (Henry Holt/Times Books, New York).

Smolin, L [2007] *The Trouble with Physics: The Rise of String Theory, The Fall of a Science, and What Comes Next* (First Mariner Books, New York).

Snow, C. P. & Collini, S. [1998] *The Two Cultures* (Cambridge University Press, Cambridge, UK).

Sokal, A. & Bricmont, J. [1998] *Fashionable Nonsense: Postmodern Intellectuals' Abuse of Science* (Picador USA, New York).

Stanford, M. [1998] *An Introduction to the Philosophy of History* (Blackwell, Malden, MA).

Stewart, I. [2002] *Does God Play Dice: The New Mathematics of Chaos* (Blackwell, Malden, MA).

Tinkham, M. [2004] *Introduction to Superconductivity* (Dover, New York).

Warnecken, H.-J. [1993] *The Fractal Company: A Revolution in Corporate Culture* (Springer, New York).

Wolpert, L. [1993] *The Unnatural Nature of Science* (Faber and Faber, London).

Wu, C. S. , Ambler, E., Hayward, R. W., Hoppes, D. D. & Hudson, R. P. [1957] "Experimental test of parity conservation in beta decay," *Phys. Rev.* **105**, 1413-1415.

Yang, C. N. [1983] *Selected Papers 1945-1980 with Commentary* (Freeman, New York).

Yorke, J. A. & Grebogi, C. (eds.) [1996] *The Impact of Chaos in Science and Society* (United Nation University Press, Tokyo).

Zhao, K.-H. [2003] "The end of the heat death theory," in *Philosophical Debates in Modern Science*, ed. Sun, X.-L. (Peking University Press, Beijing).

Zhou, T., Han X.-P. & Wang B.-H. [2008] "Towards the understanding of human dynamics," in *Science Matters: Humanities as Complex Systems*, eds. Burguete, M. & Lam, L. (World Scientific, Singapore).

Zi, C., Liu, H.-Z. & Teng, L. (eds.) [2004] *Solving the Puzzle of Competing Claims Regarding the Discovery of Parity Nonconservation: T. D. Lee Answering Questions from ScienceTimes Reporter Yang Xu-Jie and Related Materials* (Gansu Science and Technology Press, Lanzhou).

# PART I

# Art and Culture

*2*

# Culture THROUGH Science:
# A New World of Images and Stories

*Paul Caro*


The relationship between science and society is dominated by the media, which act as the middle man going in between the two. Media have their own literary set of rules which basically are those of traditional story telling recipes such as those used in fairy tales. They can be easily applied to some scientific contents as shown by examples. This practice is as old as science itself as shown by the simultaneity of the birth of modern science in the 17$^{th}$ century and its immediate popularization as texts and images by the "media" of the time. The interplay of the public opinion, shaped by the media, and the conduct of policies in which scientific knowledge is involved, or plain basic culture, is important as demonstrated by history and by contemporary conflicts involving scientific matters. Science is in fact deeply embedded in culture through words and images.

## 2.1  The Science/Society Dialogue

Scientists and experts very rarely directly dialog with the public. The scientific language is made of four components: a difficult *vocabulary*, mathematical *formulas* and tables, *numbers* (some very large, other very small), and *images* (mostly obtained through instruments which extend human senses to the immense on one side and to the smallest on the other side, and see through human body—not directly understood by people with no training in science) [Caro & Funck-Brentano, 1996]. Moreover scientific specialists from different fields do not understand each other. The dialog should then proceed from indirect ways.

In fact, the public knows about science from two sources: (1) *education*: from the elementary schools to university; (2) through the *media*: this includes the press, daily or weekly publications, the radio, the television, the books—not only learned ones but also novels including science fiction, the movies, comic stripes, and advertisements.

## 2.2 The Media in between Science and Society

Five items are discussed here.

1.  Science is another world

According to surveys of the "public understanding of science," 25% of the population can be considered as "interested" in science but only 5% are truly scientifically literate [Miller, 1994].

There is a barrier to science understanding because of the language component from the scientists: words, formulas and symbols, images, and numbers. It should be the task of education to make people more familiar with this language. But the language changes quickly and the schools, including the universities, are often lagging behind. It can be observed that a scientific language created more than a century ago is still not known and understood. This is the case for chemistry for which chemical symbols and the periodical classification of the elements are still black boxes for a majority of people. The education system of course produces the people able to deal with science in research and industry (may be also in economics?).

2.  A media-dominated society

We live in an information society, which delivers news instantly from all over the world through texts, images and sounds. The information business is part of a "show society," not only intended for plain information but also for entertainment. The excitation of the imagination of the customer is a target of media's productions.

3.  The media in between

In between the politicians, the scientists, the artists or other producers of knowledge and the public, the media act as a filter, an amplifier, a transformer, or a selection device. They shape the feelings and the attitudes of a large part of the public very often by playing with the unconscious and archetypal images.

4. Media tell stories

A newspaper article, a television show, a photograph, an advertisement, most of the time, try to tell a story. A story has to be attractive (enquiries). The rules of story-telling apply to the way the contents are selected. Politicians, scientists or artists may try to seduce the media to promote their own stories. This may be risky.

5. Story-telling recipes applied to science

An analysis of the literary tricks used by science writers (science journalists, television series, novels, etc.) show that they basically use the basic schemes of folkloric tales (the scientist as a Hero, for example, or the Monsters, or the Metamorphosis, or…).[1] Those attractive components make good stories and sustain the interest of the readers [Caro, 1996].[2] Here are some observations of interest.

*Literary tricks 1*

- The scientist as a Hero has many different figures, good or bad (the guide, the detective, the mad scientist, the Gulliver voyageur, the priest, the victim, etc.) [Haynes, 1994].
- Superhuman figures include fairies, sorcerers, good or bad devils (vitamins and dioxins), and totemic animals.

*Literary Tricks 2*

- Places, things, and times are important components of stories.
- Robots, ways of communicating at a distance, magic formulas ($E = mc^2$), numbers, universal laws, sacred places, deserts, etc.

*Mythologies*

- Metamorphosis (chemistry)

---

[1] They may also be used in economy! [Perrot, 1992].
[2] The text of [Caro, 1996] is available at: http://www.paul-caro.fr.

- Monsters (dinosaurs)
- Fatality (the entropy)
- Curses (the greenhouse effect)
- Catastrophe (earthquakes)
- Paradise lost (Mother Nature)
- Creation stories (Big Bang, Prehistory)

*A factor of confusion*

- The use of literary tricks (clichés) is common to serious reporting for a large audience and to science fiction or irrational writings (since the 1850s).
- For the public the image of science is all mixed up into those different components. The opinion is built from a mixture of rational understanding and emotional belief.

*Language*

- Most media use a poetic and metaphoric language when dealing with scientific topics.
- The French psychologist Jacques Lacan said that the language of science is basically of a mathematical nature, a rigorous language which does not permit mistakes.
- And he adds that this is the reason why so many people hate science.

*Changes*

- The mistrust is important at times where technologies are changing fast, creating and destroying activities (the 1800s, the 1920s, the 1990s).
- Then the image of the scientist and/or industrialist, as an apprentice-sorcerer is very pregnant.
- "Nature" as an all-good Deity comes back.

*Conflicts between science and law*

- Modernity is built on the simultaneous development of science, arts, and law (the Enlightenment project) to promote human happiness.
- Difficulties arise from the specialization in each of the three spheres which is translated as a lack of communication between them.
- Open conflicts are to be observed today between science and law (stem cells or genetically modified organism, GMO, for instance).

*A selective overview of science*

- Not all science topics are susceptible to yield good stories.
- Only the part of scientific activity which has some romantic overtones (research in a desert for instance) will reach the media.
- Science is then reduced within the spectacular scheme to some hot topics (astrophysics as a creation story).

*Science for very large audiences*

- Many large-audience movies have scenarios built on scientific grounds (Jurassic Park).
- From Faust to Strangelove many classical characters, mad scientist, apprentice sorcerer, are part of the plot.
- Science is associated with Power: good guys or bad guys use it.

*Science for the very young*

- The attitudes towards some sciences are shaped very early in youth.
- For instance the common image of chemistry is built on the content of comics books (Walt Disney one's for instance) [Caro, 1995].
- There the good or bad chemist, a man of power, manipulates matter for synthesis or decomposition (explosion), and always make a mistake (hence the plot).

*The scooping of science*

- Some scientists use the spectacular approach to promote their own branch of science.
- They produce best sellers by playing on the taste of people for fundamental explanations of the world or of the origins, a quasi religious urge.
- They play with a poetical language (black holes), but also manipulate obscurity (Stephen Hawkins).
- They belong to the mythical image of the "Priest," the intermediate between "worlds" in the tales (such as the King Fisher).

*The building of the image of science*

- Science is part of the culture through the way the different basic cultural media introduce it into their stories more than through pure knowledge.
- The image of science in the public is essentially the result of a combination of those different projections.
- It is often associated with *pleasure*, when *fear* is induced, it depends on the degree of personal risk perceived and changes with circumstances (the stories involving a potential risk for the *body*, such as those describing diseases, are especially sensitive to emotional responses and can even produce panics as in the case of severe pollutions with dangerous compounds like dioxins).

*The enthusiasm for science*

- Some 18th century paintings represent scientific scenes.
- For instance, from the interest in the air pump we have the somewhat cruel painting by Joseph Wright of Derby (1768) showing an experience on a bird trapped in a flask connected to an air pump (National Gallery London). An early allegory of the future conflicts of science with ethics? (There is always a large number of people watching that painting.)

*Attitudes towards science*

- The attitudes towards hot scientific issues are shaped through emotional images (for instance GMO as products of the activity of money-seeking apprentice sorcerers).
- The feeling to be at risk is strongly dependent on the strength of the imagination, more than based on an analysis of reality (compare risks of car traffic accidents versus GMO or mad cow).

*Rational and irrational*

- Surveys show that the belief in irrational in Europe is very important.
- Media support it.
- Science fiction films or novels use the same literary tricks as science reporting (the basic folkloric ones).
- The public cannot make the difference between a scientific fact and an impossible dream (teleportation and the TV series Star Trek or action at a distance with a "wave").

## 2.3 Lessons from History

Fifteen issues or observations are presented here.

1. The invention of science

Science as we know it today appears from the combination of four conditions: the application of a *mathematical framework* to the understanding of the world, the use of measurements which brings *instrumentation*, the definition of the topics to be studied as belonging to the *material natural world inert or alive*, and the creation of a *society of learned people* (supported by the State).

2. Science matters

The shift in the interest of learned people to the study of *matter* inert or alive created a practical science dedicated to produce material objects, which in the long term changed the culture and civilization, by opposition to previous times where knowledge was only *immaterial* (philosophy, theology, etc.).

3. Science in context

Since its appearance at the beginning of the 17th century, science was immediately popularized and scientific proposals were disputed in intellectual debates. Scientific principles were used to build theatrical shows [Mannoni, 1994; Stafford, 1994]. Novelists used scientific themes as literary tricks to attract interest. Politicians realized the importance of science for military purposes, economic wealth, and colonial travels.

4. An old component of the relationship between science and society

Scientists have never been isolated. Since the beginning of the 17th century their work was commented in mundane conversations, instruments were used to build attractive shows (the magic lantern 1660), and writers used scientific data in novels especially science-fiction novels (which brought an evolution of the social status of the Moon from a magical entity to a *material* body [Roos, 2001].)

5. Some examples

- *Science as part of the intellectual debate*: Thomas Hobbes (1588-1679), philosopher and
- Theoretician of the modern state; he does not believe in vacuum [Shapin & Schaffer, 1985].
- *The first science fiction novel in French*: Savinien de Cyrano de Bergerac (1619-1655): "Etats et Empires de la Lune et du Soleil," the first science fiction novel and also a popularization book [Cyrano de Bergerac, 1990].
- *The European superstar of science in the mid 17th century*: Athanasius Kircher (1601-1680), a popularizer, wrote many books; he managed one of the very first museums of science in Rome; he promoted optical shows.

6. A three-way road

Scientists publish their work for fellow scientists. Intellectuals, philosophers, and artists borrow elements from the scientist work to build their own world. Entertainment people use scientific experiments as tricks (magnets, electricity, images, etc.) to amuse crowds (in museums today).

7. The intellectual way

The presentation of science by educated people (philosophers, writers, etc.) influences the ruling class *and the media*, and may fuel the intellectual debate (Hobbes *versus* Boyle about the vacuum). The intellectual and academic world may be at time very critical of science.

8. Historical criticism of science

The bitter critic of mechanistic science and Newton at the end of the 18$^{th}$ century is part of a romantic revolt against the scientific description of the world. Science is accused of neglecting Nature at large and especially the Human Nature (Goethe, Keats, Blake, Balzac, Vigny, Nodier, etc.).

9. Sturm und Drang versus Enlightenment

In the 18th century two opposite tendencies appear which still dominate culture today. One based on rational analysis which build science and, in politics, the modern State. The other based on intuition, emotion, sensibility which dominates literature. Nature is presented as the archetype of the protective Mother on one side, as an entity to be controlled by mathematics on the other side.

10. A balance

Till the 1850s the critical romantic approach towards science and scientists dominated the public opinion. But the efforts of the scientists and of the promoters of the industrial revolution, *through the development of the press*, brought a positive coverage of science which culminated in the golden age of science popularization till the 1900s.

11. The twentieth century

Science is criticized again and this culminates in Germany after 1920 with philosopher's doubts about rationality and propaganda for the power of intuition (Oswald Spengler). A tradition of critic of science and technology is established in the intellectual world of Europe (the "children of Heidegger") [Ellul, 1988]. In the 1990s the post-modern quarrel erupted in the USA [Gross & Levitt, 1994].

## 12. A time of transition

It may be that periods of strong criticism of science are in phase with periods of creation-destruction of technologies. Today we are witnessing many components of past conflicts. The deification of Nature (the Gaia hypothesis) is one of those. The distinction between natural and artificial cannot be deconstructed scientifically; it is a social mythology, legitimate birth or not.

## 13. The weight of the media

Today the media have a strong tendency to promote and/or to create the romantic view of the world. Scientists make good targets as bad characters or visionary hero. As shown from the 19th century example, a possible strategy for the scientist is to enter the media to counteract the romantic influence by positive involvement.

## 14. How to have a voice?

Media are dominated by people with no scientific cultural background. Scientists may take steps to promote a European broadcasting system, or a wide distribution press system, which will boost *stories told by scientists*, bring science and scientists directly to the public through a conventional media system.

## 15. Stepping up

To step up really the science/society dialog it is necessary to have a scientific-community mass media approach, being able to counteract the commercial spectacular one and develop dialog. This is not impossible: many people do not trust the press but still many trust the scientists more than anyone else in society.

## References

Caro, P. [1995] "Faut-il psychanalyser la chimie?" *L'Actualité Chimique,* Avril-Mai, 5-10.
Caro, P. [1996] "Science in the media between knowledge and folklore," in *Science and the Media: The Future of Science has Begun, The Communication of Science to the Public* (Proceedings of the Vth International Conference, Milano, 15-16 February 1990), ed. Colonna, S. (Fondazione Carlo Erba) pp. 111-132.

Caro, P. & Funck-Brentano, J.-L. [1996] "L'appareil d'information sur la science et la technologie," *Rapport Commun Académie des Sciences-CADAS*, No. 6, Mai (Lavoisier Tec et Doc, Paris) p. 120.

Cyrano de Bergerac, S. [1990] "L'Autre Monde, Les Etats et Empires de la Lune et du Soleil," in *Voyages aux Pays de Nulle Part: Collection Bouquins* (Robert Laffont, Paris) pp. 277-506.

Ellul, J [1988] *Le bluff technologique* (Hachette, Paris).

Paul R. Gross, P. R. & Levitt, N. [1994] *Higher Superstition: The Academic Left and its Quarrels with Science* (Johns Hopkins University Press, Baltimore).

Haynes, R. D. [1994] *From Faust to Strangelove: Representations of the Scientist in Western Literature* (Johns Hopkins University Press, Baltimore).

Mannoni, L. [1994] *Le Grand Art de la Lumière et de l'Ombre: Archéologie du Cinéma* (Nathan Université, Paris).

Miller, J. D. [1994] "Scientific literacy: An updated conceptual and empirical review," in *O Futuro da Cultura Científica* (Instituto de Prospectiva, Lisboa) pp. 37-57.

Perrot, M.D., Rist, G. & Sabelli, F. [1992] La Mythologie Programmée: L'économie des Croyances dans la Société Moderne (PUF, Paris).

Shapin, S. & Schaffer, S. [1985] *Leviathan and the Air-Pump: Hobbes, Boyle and the Experimental Life* (Princeton University Press, Princeton, NJ).

Stafford, B. M. [1994] *Artful Science: Enlightenment, Entertainment and the Eclipse of Visual Education* (MIT Press, Cambridge, MA).

# 3

# Physiognomy in Science and Art: Properties of a Natural Body Inferred from Its Appearance

*Brigitte Hoppe*

The historical roots and development of physiognomy are presented. This chapter discusses physiognomy as a method to judge by the outer appearance of the main parts, the so-called "signs," of objects in Nature, including human beings, in order to recognize their properties. The application of this method as described is found in many scientific fields and in the arts, extending from Classical Antiquity through Renaissance Humanism, to Modern Times. In the past, both the scientist and the artist hoped to detect, by observation, the obscure—"mute" but essential—properties of a natural object. The Aristotelian legacy has been retained in popular medicine (see the doctrine of "signatures"), whereas modern sciences give preference to instrumental and experimental analytical methods. Yet from the 17th century onward, physiognomy has been used mainly to discover and represent the characteristics of a human being. The holistic approach remains a heuristic method for interpreting the complex nature of humans.

## 3.1 What Physiognomy Means and Its Methodological Aims

When observing a natural object, an animal or a human being, for the first time both the naturalist and the artist—in particular an artist of the non-abstract period—each notices the overall constitution and disposition (*habitus* in Latin) of the natural being. They ask: what is its size, its main color, the quality of its surfaces, etc. In addition, they select and describe particular features, in order to be able to recognize the object a second time. Such marks or "signs" may consist of characteristic parts of a certain size or color, etc., which are conspicuous, striking and generally present. Further, the observer assumes that these, always enduring, marks have special qualities—they may not be coincidental; they are consistent. In other words, conspicuous marks are caused by the characteristic properties of the object, are a reflection of these and represent them. Physiognomy expresses a kind of "silent knowledge" which cannot be measured precisely; it strives to uncover invisible, internal qualities on the basis of external, observable marks and signs. By using particular body marks as a guideline physiognomy aims to reveal the general disposition and character of a person, or the faculty of survival and reproduction of a natural object, as well as its usefulness for humans. This general definition was elaborated in more detail by the classical authors discussed below.

It follows then, that when the conspicuous marks or signs of a natural body are observed and described, the corresponding internal characteristics can be determined. Therefore, performing physiognomy is a method for recognizing obscure but essential qualities, and for describing them by means of words or pictures. In historical perspective, it must be emphasized that such representations were not primarily pictures of an individual body, but rather, they presented the type or abstract ideal of a collection of individuals.

In what follows we will discuss, based on the method of applied physiognomy, the extent to which natural science, medicine, and the arts (painting and sculpture) were closely connected in earlier times. Significant changes will be noted; in particular in the way that modern science no longer uses physiognomy as an analytical, but only as a heuristic method. Furthermore, we will present the continued survival of such an interpretation of natural bodies and in particular of human beings in certain disciplines and in the arts. Finally we will discuss the question of what are the common features in the world view of physiognomy, both in scientific heritage and in the arts.

## 3.2 Works of Fine Art Based on a Physiognomic Interpretation

Among the hundreds of works that exist, just a few typical examples from several periods are selected for discussion in this chapter. In

Antiquity a beard was the sign of a philosopher, as is seen in a sculpture of Plato (427-348/347 BC). In contrast, a ruler and commander like Alexander the Great of Macedonia (356-323 BC) and the Roman Emperor Augustus I (Caius Octavius, 63 BC-14 AC) is represented by a young man with curly hair. In the Middle Ages, the physiognomy found in a certain rank was often important: a ruler like Frederick II (1194-1250) is depicted sitting on a throne and his power is represented by his majestic bearing. In the Renaissance, an artist would explore certain types of emotion, say anger or rage, as expressed on the subject's face, for example in the works of Leonardo da Vinci (1452-1519) and others [Caroli, 1995; Reisser, 1997]. In 17$^{th}$ and 18$^{th}$ century portrait painting, a particular sign on the face of a man or woman indicated a certain characteristic; for example a high forehead was interpreted as an indicator of intelligence. In the period of classical modern painting around 1900, especially in expressionist painting, artists tried to show deep inner emotions. The masterpieces of the Norwegian painter Edvard Munch (1863-1944), such as his "Crying Human on a Bridge," provide eminent examples.

## 3.3  Physiognomy in Science

Three major items will be presented here.

### 3.3.1  *Early Roots of Physiognomic Practice*

The beginnings of the practice of physiognomy were linked with mantic practices in the ancient Egyptian and Babylonian societies [Touwaide, 2000]. It is also thought to have been practiced by the ancient Greeks from very early date, judging by an anecdote which tells of Socrates being confronted once by the physiognomist Zopyros at the agora of Athens, where interpreters of dreams, astrologists, and chiromantists peddled their skills [Lavater, 1781: Part 1; Vogt, 2001; Baumbach, 2002]. Physiognomy was applied in particular in Hippocratic medicine, in certain parts of Empedoclean philosophy, and in the works of poets such as Anakreon and Aischylos.

## 3.3.2 The Fundamental Treatise of Aristotle and Its Legacy

The first special scientific treatise defining more precisely the practice of physiognomy and explaining its efficacy by means of natural philosophy is ascribed to Aristotle, since 1988 no longer considered a pseudo-work [Degkwitz, 1988]. In the treatise known as *Physiognomoniká* the possibility for recognizing the essential nature of an animal or a human being by external marks is based on the Aristotelian doctrine of the soul (psychē). Because all external qualities of a natural body were closely connected with the invisible qualities of the soul of the individual, it seemed possible to interpret the observable marks as signatures indicating the qualities of the essential nature of an animal or a human being. Such close connections became evident to Aristotle; distinct changes in the state of a disease or intoxication of an individual were observable.

The next section discusses the importance of the early analytical doctrine of physiognomy to the development of natural sciences. Becoming the rank of a discipline by the Aristotelian treatise, physiognomy's aim was to introduce some standards or signatures for characterizing the properties of an animal or a human being, in order to discern some general types. The Aristotelian text listed the marks or "signs" to be observed as a basis for discovering their essential nature. Certain main sentences of Aristotle's text are quoted here:

ARISTOTLE: Physiognōmoniká [Bekker, 1851: 806 a-b]

The physiognomonía (φυσιογνωμονία) points out [...] the natural qualities (τὰ φυσικά παθήματα, naturales passiones) of the characteristics of the soul and those acquired qualities which alter, if they become manifest, the signs (τῶν σημείων, signa), which the physiognomonia observes. [...]

The physiognomonía observes the motions, the figures (ἐκ τῶν σχημάτων, ex figuris), the colors, the facial expressions, the quality of the hairs and of the surface, the voice, the flesh, the parts of the body and the entire stature. [...]

After the general explanation in the first part of the treatise, Aristotle characterized in the second one several well-known animals such as the lion, a neat, a dog, and a number of birds. He compared the structure of certain body marks, such as thick or fine hairs and a deep or low voice of different animals, generalized the marks as signatures of types, and then adopted the system to human beings, in order to characterize their mental qualities. The first group of marks indicated the signature of a courageous animate being, while the second group indicated a timid or faint-hearted one. Many physical qualities should be related to both sexes in the following manner:

ARISTOTLE: *Physiognōmoniká* [Bekker, 1851: 806 b]

> The figures (τὰ σχήματα, *figurae*) and the facial expressions (τὰ παθήματα τὰ ἐπιφαινόμενα, *passiones* [...] *in facie*) are selected namely in accordance with the similarity of the emotions, together with which they appear. [...] The male [type] is higher than the female, its limbs are stronger, more flexible, by means of its stature and quality more suitable for all virtues (τὰς ἀρετάς).

Different types of animals or humans could be discerned quite clearly by means of the Aristotelian catalogue of signatures. Therefore, the close connection between philosophic presuppositions and evident experiences seemed to present a plausible doctrine. This Aristotelian doctrine was documented and communicated by a long tradition of texts dealing with physiognomy, from Antiquity to the 19th century (Table 3.1).

After a long period of a more or less uncritical tradition of the main contents and the practice of physiognomy, its philosophical basis was explained in detail during the period of Humanism once again [Hoppe, 1998]; for example the Italian humanist Alessandro Achillini in 1503 and his disciple Bartolomeo Cocles in 1504 emphasized the character of physiognomy as a doctrine, a science and a special discipline, although they classified it together with the chiromancy as a discipline subordinate to natural philosophy. Later in the 18th and 19th centuries, the connection with Aristotelian philosophy was forgotten, but empirical practices remained. Some of these are still applied being at the present time; however, now they are often linked with other empirical methods.

Table 3.1. The early tradition of main authors and texts about physiognomy.

| 13$^{th}$ c. | Albertus Magnus, Thomas Aquinas |
|---|---|
| 12$^{th}$ c., 2$^{nd}$ part | Fakhr al-Dīn al-Rāzī: al-qiyāfa, al-firāsa |
| 7-9$^{th}$ c. | *Secreti Secretorum Pseudoaristotelici,* — *De Physiognomonia* (lat.), *Capitulum: Homo* — |
| 5-6$^{th}$ c. | *Anonymi De physiognomonia liber* (lat.) (concerning humans and animals) |
| ca. 400 AC | Adamantios, Alexandreia: *Physiognōmoniká* |
| ca. 90-145 AC | Antonius Polemon, from Laodikeia, Sophist: *Physiognōmoniká* (humans, animals) (lat., *De physiognomia liber*; Arabic translation in 1356) |
| 4 (- 3$^{th}$) c. BC | Aristotle: *Physiognōmoniká* (humans, animals) |
| 4$^{th}$ c. BC | Aristotle: *Analyt. prior., Hist. anim.*, etc. (animals, humans) |
| 6-5$^{th}$ c. BC | physiognomic fragments concerning human beings in texts of Greek poets, Empedocles, and Hippocratic medicine |

Although after the end of the Middle Ages, the practice and the doctrine of physiognomy was sometimes criticized for example by Agrippa von Nettesheim, the Italian Giambattista della Porta, and the critical German philosopher and physicist Georg Christoph Lichtenberg, the doctrine of physiognomy continued to be passed on and applied for characterizing not only animals and human beings, but also other natural objects.

### 3.3.3 The Impact of Aristotelian Physiognomy in Natural Science

1. Physiognomy of plants

In past times botany physiognomic interpretations of plant qualities had an important theoretical significance, which was understood neither by historians of botany, nor by philologists of the classical languages. The physiognomic ideas helped the naturalists to discern very similar-looking, but not identical species of plants long before they had been able to describe precisely enough all essential organs and the physiological functions of higher plants. In particular, the often miniscule reproductive organs and their functions remained unknown, although the production of fruits and seeds as well as their faculty to produce new individuals were observed. This property of fertility and the widely used method of thinking by means of analogies led the early founders of scientific botany to interpret some physical marks of a plant like those of an animal as sexual signatures. In the Aristotelian text about physiognomy, the following opposing qualities of animal bodies were declared as typical for both sexes (Table 3.2):

Table 3.2. Observable marks typical for a male and female animal.

| Mark | Male type | Female type |
| --- | --- | --- |
| overall stature | large-sized | small-sized |
| single parts | strong, hard, thick | slender, soft, tender, thin |
| size | angular | round |
| surface | rough | smooth |
| color | deep, dark | light, pale |
| odor | fragrant | less fragrant, disagreeable |
| voice | deep | small |
| reproduction | "sterile", fruitless | fertile, fruitful |

Different marks having such qualities were used by Theophrastos (ca. 372-287 BC), the founder of scientific botany, in order to classify plants, mainly trees. His main successors, such as Pliny the elder (23/24-79) in Roman Antiquity, Albert the Great in the Middle Ages (ca. 1200-1280), and especially humanists such as the German physician and botanist Leonhart Fuchs (1501-1566) in 1542 and 1543 continued in this tradition [Hoppe, 1998] (Table 3.3).

Table 3.3. Physiognomy in the description of plants.

| Theophrastos[1] | Translation[2] | Fuchs[3] |
|---|---|---|
| "Philyra altera est <u>mas</u>, altera <u>femina</u>: differunt tum tota forma, tum ligni, et quod hæc fructifera, altera sterilis est. Maris <u>lignum</u> est durum, flavum, nodosius et densius, etiam euodoratius, feminæ autem magis album; atque <u>cortex</u> maris crassior et detractus parum flexibilis propter duritiem, feminæ vero tenuior et flexibilis, e quo cistas faciunt. Mas sterilis est et flore caret, femina vero et florem et fructum fert: <u>flos</u> calyciformis [...] viridis dum in calice est, apertus autem flavescens. <u>Folium</u> [...] figura hederaceo simile [...]" | The lime-tree: one is the <u>male</u>, the other one the <u>female</u>; both sexes differ from one another according to the overall stature, to the wood and in so far as one is fertile, the other one is sterile. The <u>wood</u> of the male is hard, yellow-red, more knotty and solid, more fragrant too; that of the female is more white. The <u>bark</u> of the male is thicker and, pulled off, less flexible because of the hardness; but that of the female is thinner and flexible, and used for making boxes. The male is sterile without a blossom; the female produces a blossom and fruit. The cup-shaped <u>flower</u> [...] is green so long as it remains in the calyx, but unfolded it looks pale yellow. The <u>leaf</u> is shaped resembling that of ivy [...] | "Antique authorities discern two sexes of the lime-tree, the <u>male</u> and the <u>female</u>. The male has hard, yellow, knotty <u>wood</u> [...]. The female has a white wood [...]. It produces a blossom and fruit. The <u>flowers</u>, as long as sticking in the calyx ("in their sacs"), are green, if unfolded, they look pale yellow. Its <u>leaves</u> look like the foliage of ivy [...]" |

[1]Theophr. HP 3.10.4. [2]By Brigitte Hoppe (cf. Hort [1961]). [3,2][Fuchs, 1543] Chap. 332.

Moreover, Fuchs adopted the method, in order to describe plants, which had not been included in the botanic writings of Theophrastos [Hort, 1961] or others like the wall-pepper or wall-grass species, one smaller, with white flowers, called the female, and the other one with darker, yellow-red flowers, the male (*Sedum album* L. and *S. rupestre* L., subsp. *reflexum* Hegi et Schmid, *Crassulaceae*). The physiognomic interpretation was applied even on lower plants which never produce flowers and fruits. The Italian botanist Pietro Andrea Mattioli (1501-1577), following Dioskurides (ca. 70 AC),[1] called a species of ferns with a black rhizoma the male ("Filix Mas"; to identify with the male fern, *Dryopteris Filix-mas* (L.) Schott, *Polypodiaceae*) (Fig. 3.1) and the other one with a lighter rhizoma showing a yellow gleam the female ("Filix Foemina"; to identify with the bracken, *Pteridium aquilinum* (L.) Kuhn, *Polypodiaceae*) (Fig. 3.2).

---

[1] See [Mazal, 1998] on Dioskurides [*Mat. Med.*], Book IV, Chap. 184 and Chap. 185.

NOMINA.                                              991
Filix, Græcis, Πτέεις. Arabibus, Sarax seu Sarachs. Italis, Felce. Germ. Waldt-
farn. Hispanis, Helecho jerua. Gallis, Osmunda Regale, seu Fengiere masse.

FILIX.

GENERA.

Duo eius habétur genera, Diose.
Mas & Fœmina.

FORMA MARIS.

Filici folia, sine caule, sine Idem.
fructu, sine flore ex vno pedi-
culo cubitali longitudine ex-
eunt, multifida, è laterib. pri-
uata, & expansa velut ala, sub-
graui odore. Radice per summa
ma cespitum nigra, oblonga,
filices multos fundente, sub-
stringenti gustu.

LOCVS.

Nascitur in montibus, & Idem.
saxosis. Quin etiam alibi vm Matth.
brosis locis.

QVALITATES.

Mas radicem habet maxi- Gal. li. 8.
mè vtilem: latum enim lum- simp. me.
bricum interficit. Est enim a-
mara, paulum habens adstrictionis, valentem desiccandi facultatem obtinet,
nec tamen mordax est.

Fig. 3.1 Mattioli, P. A. [1586]: The chapter on "Filix Mas", the malefern (*Dryopteris-Filix-mas* (L.) Schott, *Polypodiaceae*) [Photography by the Deutsches Museum, München.]

The identification of those plants by means of the modern systematic botanical categories shows us that Fuchs distinguished species of the same genus, but not dioecious plants, that is, really sexually differentiated individuals; moreover, it shows that Mattioli described sexless individuals of cryptogamic plants; therefore, we may argue that the impact of physiognomic ideas in botany was not important for its scientific development. But even the "misinterpretations" had a historic role and are remarkable: This relatively superficial interpretation of vegetative, morphological marks as signatures of sexual differentiation led to the search for the real sexual organs and the acceptance of the new theory of plant sexuality from the time of Linnaeus until the 19th century.

Fig. 3.2. Mattioli, P. A. [1586]: The chapter on "Filix Foemina", the bracken (*Pteridium aquilinum* (L.) Kuhn, *Polypodiaceae*) [Photography by the Deutsches Museum, München.]

## 2. Physiognomy in medicine—the doctrine of signatures of natural objects

A remarkable broadening of physiognomic interpretations began in the Middle Ages and continued throughout the Renaissance. Several Moslem scholars (to our present knowledge) dealt with physiognomy in long treatises, one of the main authors was Fakhr al-Dīn al-Rāzī who wrote on "al-firāsa"; that is, the method for deducing internal qualities from external marks of an object, at the end of the 12$^{th}$ century († 1210). He included not only animals but also plants and minerals in his physiognomic interpretations. This tradition was continued by humanistic authors from around 1500 onward. While Alessandro Achillini, the Italian humanistic anatomist working in Bologna, emphasized the importance of human physiognomy and chiromancy in 1503, his disciple, the master Cocles or Bartolomeo della Rocca, extended this to include also animals, plants, stones and metals in a voluminous work in 1504. In one chapter he treated a new area: the "*physiognomonia planetarum*", the physiognomy of planets. He explained the physiognomic meaning of main marks of the seven planets and of some of the signs of the zodiac, such as the size, brightness, clearness, brilliancy, and "color" (from white to red), which are meant to also reflect certain sexual qualities.

The humanistic authors accepted Aristotelian natural philosophy and added elements of other philosophies; for example the Neoplatonic doctrine of emanation and the idea that all parts and bodies of the cosmos should be connected with one another by sympathy. The invisible connections between different natural bodies and their effects on the human body are to be detected by the physiognomic method. Because the scholars of the 16$^{th}$ and 17$^{th}$ centuries were convinced of sympathetic connections between natural bodies, which meant that occult qualities could be observed by means of external marks, they developed the physiognomic method systematically into the doctrine of "signatures". One of the most fruitful authors in this field, Giambattista della Porta of Naples († 1615) emphasized that the doctrine was based on "natural reasons" (in Latin *physicae rationes*) [Hoppe, 1998]. The doctrine became the basis of therapeutics from the Middle Ages onward, a healing system which was widely propagated in the early part of the modern period, and has carried through, though only in "popular" medical treatment, until the present time.

The system of "signatures" influenced the recipes and even the names and plant illustrations in many herbals since the Middle Ages. In an illustrated version of the 14th century of the influential medieval pharmacopoeia with the *incipit* "*Circa instans [...]*" we find a physiognomic picture of the herb called "pulmonary herb" (it is called in modern botanical nomenclature *Pulmonaria officinalis* L., *Borraginaceae*). It shows only two large leaves without the stem and flowers. They are arranged like the wings of the human lungs; and broad white spots seemed to resemble the alveolar structure of the lungs. These marks of the leaves of the pulmonary herb can be observed; they seemed to indicate their faculty to heal bronchial and lungs afflictions. Such indications of healing virtues of plants and minerals were based on similar interpretations by many authors; for example Paracelsus (1493/94-1541) mentioned them in his medical works (Table 3.4).

The learned Italian philosopher Giambattista della Porta (1539–1615) examined the main fields of physiognomy in special monographs at the end of the 16th century. Based on the traditional texts of Aristotle and of Polemon from the first half of the second century after Christ, Porta illustrated his book on human physiognomy in 1586 with pictures comparing human faces with animal ones. In his *Phytognomonica* of 1588 he compared parts of plants with human organs such as the heart or liver, and also with small animals. The root of a composite herb looked like a scorpion; therefore, this root should be a remedy against the poison of a scorpion. Moreover, Porta wrote *Physiognomiæ Cælestis Libri Sex* in 1601, which dealt with the physiognomy of planets and the signs of the zodiac, since Porta rejected their astrological relation to human beings. This same aim, to show the "signatures" of natural objects, was the topic of a book by Oswald Croll (ca. 1560-1609) in 1608 and of a treatise by the German chemist, Johann Rudolph Glauber (1604-1670) in 1658. The latter author recommended particular salts, minerals and metals based on observable qualities such as their color, solidity, solubility, flavor, etc., as remedies with different indications. The doctrine of "signatures" was so highly esteemed, mainly in the field of medicine from the 16$^{th}$ to the 18$^{th}$ centuries, that an immense quantity of treatises was published in Latin and in various European languages [Hoppe, 2002]. Though this tradition of the so-called phyto-medicine has been repressed by chemotherapy since the end of the 19$^{th}$ century, it has remained a strong element of popular medicine. One of the most recent books *Nature and Signature of the Healing Herbs* was published in Switzerland in 2002 by Roger Kalbermatten (AT-Verlag, Aarau), a popular author.

Table 3.4. Tradition of authors and texts about physiognomy from Renaissance onward.

| | |
|---|---|
| 1777 | Georg Christoph Lichtenberg: *Über Physiognomik* |
| 1775–1778 | Johann Caspar Lavater: *Physiognomische Fragmente [...]* |
| 1658 | Johann Rudolph Glauber: *Tractatus de Signatura Salium, Metallorum et Planetarum* |
| 1647, $^2$1659 | *Anatomia et Physiognomia Simplicium* (J. Gudrio de Tours) |
| 1625 | Rudolphus Goclenius: *Physiognomia et Chiromantica [...]* |
| 1608 | Oswald Croll: *Tractatus de Signaturis internis Rerum seu de [...] Anatomia maj[oris] et min[oris] Mundi* |
| 1601 | Giambattista della Porta: *Physiognomoniæ Cælestis Libri Sex* |
| 1588 | Giambattista della Porta: *Phytognomonica* |
| 1586 | Giambattista della Porta: *De humana physiognomonia* |
| 1550 | Antoine du Moulin: *Fisionomia [...]* |
| ca. 1520–40 | Paracelsus (Theophrastus Bombastus von Hohenheim) *De natura rerum [...]* |
| 1530 | Agrippa, in: *De incertitudine et vanitate scientiarum* |
| 1510 | Agrippa von Nettesheim, in: *De occulta philosophia* |
| 1504 | Bartolomeo Cocles: *Chyromantie ac Physionomie Anastasis* |
| 1503 | Alessandro Achillini: *De Chyromantiæ principiis et Physionomiæ* |

### 3.3.4 The Renewal of Physiognomy for Characterizing a Human Being

1. Intense application of physiognomy in medicine and psychology in the 18th and 19th centuries

For characterizing humans based on the size and countenance of the face and how it changes under the influence of different emotions, the Swiss theologian Johann Caspar Lavater (1741–1801) once again compared humans with animals in his voluminous, illustrated publication from 1775 to 1778. Moreover, Lavater underlined that physiognomy should be developed, in order to establish certain principles of this science. He introduced a geometric network including the main parts of human's face profile for comparing different distances and figures between remarkable points, such as the hairline, eyebrow, bridge and tip of the nose, lips, and chin (Fig. 3.3). He supposed that these lines may indicate important individual properties of the human being, but he conceded that he was judging human's character by means of "sentiment and experience". The profile "a" of a young man (see Fig. 3.3) represented a consistent, happy, and firm character, a man without boldness, but with an inclination to sensitivity which he knew to control. To the man with the profile "f" Lavater ascribed a good sense, prudence, sincerity, and love of order, but a flaw of the faculty of judgement. Throughout several decades his speculations were discussed and criticized at the same time [Borrmann, 1994; Percival & Tytler, 2005].

The increase in anatomical knowledge in the 18th and 19th centuries carried a much greater impact on science and medicine. Human skulls were examined and studied by Albrecht von Haller (1708-1777) among others, and were the subject of comparative anatomy in the works of Pieter Camper (1722-1789), Georges Cuvier (1769-1832) and Johann Friedrich Blumenbach (1752-1840) later on. By means of interpretation of intensive observations about the functions of senses, nerves and their connections with the brain, several French philosophers and professors of medicine such as Charles Bonnet (1720-1799), Pierre Jean George Cabanis (1757-1808), and Philippe Pinel (1755-1826) strove to understand changes in bodily marks caused by diseases. The nervous systems,

senses and the brain were studied by the comparative anatomist Samuel
Thomas Soemmerring (1755-1830) and by Franz Joseph Gall (1758-
1828). The later scientist in particular began to found a theory on the
localization of emotions, and intellectual and psychic faculties in
different parts of the brain [Gall, 1810-1819]. Later on, he and even more,
his disciple Johann Christoph Spurzheim (1776-1832), established on the
basis of morphologic-anatomical studies the hypothesis of cranioscopy in
1815, called phrenology later on [Spurzheim, 1815]. This practice
attempted to establish a connection between the figures of external parts
of the skull as "signatures" of the intellectual and mental faculties of a
human. Phrenology was discussed and applied in Europe and in the
United States during the 19th century [Hartley, 2001].

The efforts to re-introduce physiognomic ideas did motivate physicians to make more accurate observations of physical changes in the
pathological cases, they were treating. Thus they were known to apply
their acute observations as a means of improving their diagnostics. Of
particular note, among others, is the friend of J.W. von Goethe (1749-
1832) Carl Gustav Carus (1779-1869), and the medicine professor at
Freiburg im Breisgau (Black Forest), Karl Heinrich Baumgaertner (1798-
1886), who published a book entitled *Krankenphysiognomik* (*The
Physiognomy of the Ill*) in 1839. One of the most popular authors of the
time was Carl Huter (1861-1912), who discerned three main human
types characterized by their body marks. He propagated by means of his
healing concept from 1894, called "Psychophysiognomik" (Psycho-
physiognomy), a reformation of the entire popular lifestyle, with the aim
of improving the human character with the aid of proper diet and
physical training [Borrmann, 1994]. His concept has had many followers
[Tepperwein, 2003]. The physiognomic practices were applied not only
in Western medicine but also in Chinese traditional medicine [Bridges,
2005]. By including many additions following a multitude of further
empirical observations, physiognomy has remained a helpful tool for
medical diagnostics until the present time.

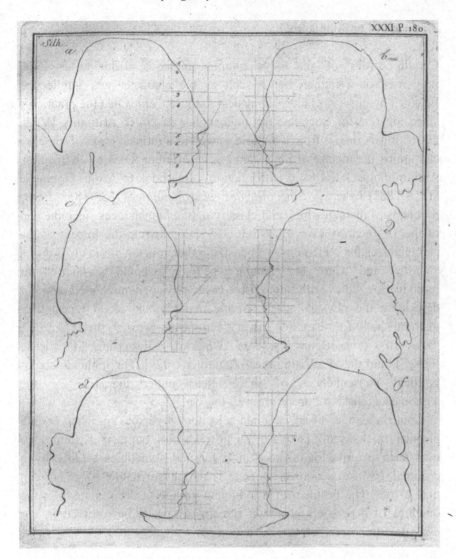

Fig. 3.3. Lavater, J. C. [II, 1783, see p. 180]: Six different profiles with lines for characterizing important parts of the face [Photography by the Deutsches Museum, München.]

## 2. Misuse of physiognomy

Certain critics of the physiognomic methods deducing essential properties of a human being such as the character and intellectual faculties by observing several physical marks accompanied the practicing since the time of Socrates, the philosopher of Greek Antiquity [Vogt, 2001]. Furthermore, from a historic as well as a rational point of view we can notice the misuse of physiognomy in different fields of civilization. Anthropologists discussed with relation to the book of Arthur Gobineau (1816-1882) *Essai sur l'inégalité des races humaines* (1853-1855, Paris) problems of distinguishing and classifying the human races since the end of the 19th century [von Eickstedt, 1937]. Following the impact of the Nationalsocialist's ideology, emphasizing the eminent ethnic quality of a "Nordic" and "Germanic" human type which should be favored for reproduction, the anthropologists used physiognomic methods for describing the physical marks of the main human racial types; they distinguished for example five "European races": the "Nordic", "Western" ("westische Rasse"), "Dinaric", "Easterling" ("ostische Rasse"), and the "Eastbaltic race" [Günther, 1929]. The inhuman socio-political consequences of these ideas under the rule of the Nationalsocialists in Europe are well-known.

But the danger of an inhuman selection of persons by using certain physical marks is still lasting in social fields until our days. Criminology uses such external marks to search for and sometimes to identify a criminal. Even in school or work tests the physiognomic judgment can play a role. The problem of a judgment going not deep enough on the qualities of a person's character remains a task to be resolved in our everyday life.

### 3.3.5 *The Physiognomy of Vegetation Characterizing a Landscape*

At the same time as comparative methods were being widely applied in zoology and botany to discover typical morphological structures of the main organs and bodies of animals and plants, the eminent naturalist, Alexander von Humboldt (1769-1859), introduced the notion of physiognomy to the field of geography. He had seen many very different regions of the globe during his long expeditions in several countries of the Old and the New World. Humboldt was inspired by landscape painting which had been flourishing since the late 17th and the 18th centuries. He characterized different landscapes by describing by means of words their conspicuous marks such as a grouping of rocks or trees of different sizes, which constituted a "natural painting" ("ein Naturgemälde"). He sought out physiognomic types, not only in human faces and on the bodies of animals and plants, but also in their natural environment. This was especially helpful for understanding the relationship between a plant species or plant society, relatively fixed at certain localities, on the one hand, and geographic and climatic regions on the other. In order to characterize diverse types of landscapes, Humboldt used the frequent occurrence of very conspicuous plant genera or species in a particular area. In his book *Ideas for a Geography of Plants,* from 1805/1806 he described 17 different plant types such as palm-trees, orchids, grasses, lichens, etc. In every instance, their special combinations characterized a region or landscape. Single types of vegetation could indicate the chemical and physical state of the soil as well as the qualities of the atmosphere and of the climate in a special region. They indicated, like "signatures" in a sense, a whole set of environmental factors, which could not be ascertained only by observation, but more by precise measuring and experimental research.

The physiognomic interpretations made by Humboldt were accepted immediately by nearly all German-speaking plant geographers and by a number of French colleagues, with the exception of Alphonse de Candolle (writing 50 years later, in 1855), and by various botanists at the end of the 19th century [Hoppe, 1990]. Humboldt's views have been applied in the field of ecology until the present. As for example cultivated forests in certain alpine regions of Europe are constituted only by a species of spruces (*Picea* species, *Pinaceae*), the ecologists name the forest type from this plant genus. Thus, a certain type of biotope in a particular region or landscape is characterized by a plant group dominating the vegetation.

## 3.4 Conclusion

Let us conclude with a number of observations and points of interest:

1.  The long and multi-faceted tradition of physiognomy in science and art indicates the broad degree of acceptance that this interpretation of natural objects has enjoyed. Scientists and artists tried to discover and to represent the typical characteristics of their objects of observation. Both the scientist and the artist strove to detect obscure, "mute", yet essential properties of a natural object. The physiognomic approach seemed to recognize not only material properties of a natural body, but also the mental faculties of a living being ascribed in earlier times, to higher animals as well as to human beings.

2.  In the modern arts, the aims have changed: a portrait presents the individual features; certain symbols or pure figures and colors, when using abstract methods, convey a message. Other particular physiognomic features are found mainly in caricatures (see the drawings of the Norwegian painter Edvard Munch: showing dogs and a human being in a similar posture; Fig. 3.4).

3.  The modern biological and chemical sciences prefer methods of measuring by means of instruments and experimental methods for analyzing and determining the characteristics of natural bodies. The holistic approach for characterizing a landscape is preserved as a heuristic method in ecology.

4.  From a historic point of view we have to notice also the possibility of misuse of physiognomy in different historic contexts. Anthropologists began to use physiognomic methods with the aim to determine mental faculties of human beings by accurate observation and "exact" measurement of physical marks from the end of the 19th century onward. Thus, this tradition was misused under the impact of the ideology of Nationalsocialism which practiced ethnic selection of human beings. The problem of an easygoing judgment for identifying and selecting persons exists also in the social life. Observing human's physiognomy in school tests or criminal anthropology can produce wrong decisions.

5. In those disciplines which deal directly with human beings, physiognomy has been preserved, at least as a heuristic method: that is, in medical diagnostics, psychology, behaviour physiology, and popular medicine. These areas are confronted with obscure properties of a human being which are not measurable in their entirety. The complex nature of a human being retains a certain degree of mystery until the present day.

Fig. 3.4. Edvard Munch: Drawings showing dogs and a human being in similar postures (from Catalogue No. 44 in [Hougen, 1970]). [Photography by the Deutsches Museum, München.]

# References

Baumbach, M. [2002] "Zopyros [3]," in *Der Neue Pauly. Enzyklopädie der Antike, Altertum,* Vol. **12/2**, eds. Cancik, H. & Schneider, H. (J. B. Metzler, Stuttgart, Weimar) col. 835.

Bekker, I. (ed.) [1851] *Aristotelis Opera,* Vol. **2** (G. Reimer, Berlin).

Borrmann, N. [1994] *Kunst und Physiognomik* (DuMont Buchverlag, Köln).

Bridges, L. [2005] *Gesichtsdiagnose in der chinesischen Medizin,* German Transl. of *Face Reading in Chinese Medicine* (Elsevier, Urban & Fischer, München).

Caroli, F. [1995] *Storia della fisiognomica: Arte e psicologia da Leonardo a Freud* (Mondadori, Milano).

Degkwitz, A. [1988] *Die pseudoaristotelischen "Physiognomonica" Traktat A. Übersetzung und Kommentar* (Phil. Diss. Univ. Freiburg im Breigau).

Fuchs, L. [1543] *New Kreüterbuch/in welchem nit allein die gantz histori/das ist/namen/gestalt/statt vnd zeit der wachsung [...]* (Isingrin, Basel).

Gall, F.J. [1810-1819] *Anatomie et physiologie du système nerveux en général, et du cerveau en particulier, avec des observations sur la possibilité de reconnoitre plusieurs dispositions intellectuelles et morales de l'homme et des animaux, par la configuration de leurs têtes* (F. Schoell, Paris).

Guenther, H. F. K. [1929] *Rassenkunde Europas* (Lehmann, München).

Hartley, L. [2001] *Physiognomy and the Meaning of Expression in Nineteenth Century Culture* (*Cambridge Studies in Nineteenth Century Literature and Culture,* **29**) (Cambridge University Press, Cambridge).

Hoppe, B. [1990] "Physiognomik der Vegetation zur Zeit von Alexander von Humboldt," in *Alexander von Humboldt. Weltbild und Wirkung auf die Wissenschaften,* (*Bayreuther Historische Kolloquien,* **4**) (Böhlau, Köln, Wien) pp. 77-102.

Hoppe, B. [1998] "Physiognomie der Naturgegenstände insbesondere der Pflanzen in der Antike und ihre Wirkung," in *Antike Naturwissenschaft und ihre Rezeption* **8**, 43-59.

Hoppe, B. [2002] "Physiognomik III: Zoologie, Botanik, Mineralogie," in *Der Neue Pauly. Enzyklopädie der Antike, Rezeptions- und Wissenschaftsgeschichte,* Vol. **15/2**, ed. Landfester, M (J. B. Metzler, Stuttgart, Weimar) col. 358-362.

Hort, A. (ed.) [1961] *Theophrastus, Enquiry into Plants [Historia Plantarum],* Vol. **1** (*The Loeb Classical Library*) (W. Heinemann, London).

Hougen, P. [1970] *Edvard Munch. Das zeichnerische Werk, Katalog der Ausstellung der Staatlichen Graphischen Sammlung München* (H. M. Hauschild GmbH, Bremen).

Lavater, J. C. [1781-1787] *Essai sur la Physiognomonie, [...],* Partie **1-4** (Jacques van Karnebeek, La Haye).

Mattioli, P. A. [1586] *De Plantis Epitome [...]* (Feyerabend, Frankfurt am Main).

Mazal, O. (ed.) [1998] *Der Wiener Dioskurides [Materia Medica]*, Vol. 1–2 (Akademische Druck- und Verlagsanstalt, Graz).

Percival, M. & Tytler, G. (eds.) [2005] *Physiognomy in Profile: Lavater's Impact on European Culture* (University of Delaware Press, Newark).

Reisser, U. [1997] *Physiognomik und Ausdruckstheorie der Renaissance* (*Beiträge zur Kunstwissenschaft*, **69**) (Scaneg Verlag, München).

Spurzheim, J.C. [1815] *The Physiognomical System of Drs. Gall and Spurzheim: Founded on an Anatomical and Physiological Examination of the Nervous System in General, and of the Brain in Particular; and Indicating the Dispositions and Manifestations of the Mind* (Baldwin, Cradock, and Joy, London).

Tepperwein, K. [2003] *Krankheiten aus dem Gesicht erkennen* (Weltbild Verlag, Augsburg).

Touwaide, A. (Transl.: Heinze, T.) [2000] "Physiognomik," in *Der Neue Pauly. Enzyklopädie der Antike, Altertum*, Vol. **9**, eds. Cancik, H. & Schneider, H. (J. B. Metzler, Stuttgart, Weimar), col. 997-998.

Vogt, S. [2001] "'Wir urteilen stündlich aus dem Gesicht, und irren stündlich', Perspektiven der Physiognomik," *Einsichten, Forschung an der Ludwig-Maximilians-Universität München* **19**, 38-41.

Von Eickstedt, E. [1937] *Rassenkunde und Rassengeschichte der Menschheit*, Vol. 1 (F. Enke, Stuttgart).

# 4

# Has Neuroscience Any Theological Consequence?

*Alfredo Dinis*

In Michael Persinger's book *Neuropsychological Basis of Human Belief* (1987), he claimed to have found a causal correlation between the frequency of epileptic seizures affecting the left temporal lobe of human beings, and the frequency of their religious experiences. Rhawn Joseph has argued along the same lines in *The Transmitter to Go: The Limbic System, the Soul, and Spirituality* (2000). More recently, research on the neural correlate of religious experiences, such as Buddhist and Christian meditation, have become an interesting line of research for neuroscientists. The fact that a neural correlation of every religious experience can be identified, has led some authors to put forward the thesis that religion is fully explainable in terms of neural activity. I will try, first, to clarify whether we are allowed to identify the neural correlation of every human experience (its necessary condition) with its causation (or sufficient condition). Secondly, I will discuss the difference between internalist and externalist approaches in the study of the mind in general, and of the neural correlate of religious experiences, in particular. And thirdly, I will attempt to understand whether neuroscience fully explains the complexity of religious experience.

## 4.1 Neurotheology

How religious experiences correlate to the activation of specific brain areas is a matter of intense debate. Some neuroscientists claim that such experiences are deterministically caused by the activation of some brain areas.

Michael Persinger is certainly one of the best known authors of this group. Since the publication of his book *Neuropsychological Basis of Human Belief*, Persinger has been claiming that religious experiences—

such as feeling at one with the universe, having visions, hearing voices, etc.—are just a matter of neural activation either by epileptic seizures or through laboratorial stimulation of parts of the left temporal lobe, as performed by Persinger himself at the Laurentian University in Ontario, Canada [Persinger, 1987]. Participants wearing special helmets called transcranial magnetic stimulator had their temporal lobes stimulated by weak magnetic fields. Most of them reported they had felt a non physical presence in the room. Persinger consider that this sense of a felt presence is the prototype of a religious experience, but I will argue later on that this is a highly disputable view. However, he claims to have enough experimental evidence that religious experience can be simulated in the laboratory [Persinger, 2003].

For Persinger, god beliefs consist of two different elements: the god *experience* and the god *concept*. The god concept is associated with the linguistic concepts of each culture, and explains the many names that are given to gods, spirits and other supernatural entities and experiences. It is the god experience that is associated with the activation, or lack of it, of some specific areas of the brain. Such areas are related to the limbic system, which explains perfectly why religious experiences are basically emotional. However, the fact that any human experience, religious or not, has some sort of correlation with brain activation does not necessarily mean that such a correlation fully and causally explains human experiences. Moreover, Persinger oversimplifies that correlation. It is a well-known fact that for every experience several brain areas are synchronically involved. It is also known that not everybody that suffers from epileptic seizures reports religious experiences. But it is also a fact that not everybody that goes through electromagnetic stimulation of the temporal lobe reports any religious experience. Persinger performed such an experiment in his laboratory with some volunteers who reported what he interpreted as religious experiences [Persinger, 2003]. However, he applied to the well-known atheist Richard Dawkins the same technique he has applied to several other people who, as a consequence, reported they went through religious experiences. The laboratorial session with Dawkins lasted 40 minutes. In the end, he said he felt that the magnetic fields produced around his temporal lobes affected his breathing and his limbs, but did not produce any religious experiences. Persinger was not impressed by this failure to find a necessary connection between brain activation of specific brain areas and the production a religious response. He argued that Dawkins was not sensitive enough to the stimulation of the temporal lobe by magnetic fields, so he could not feel the same effects of other people, an ad hoc explanation that I find quite unsatisfactory. In fact, Dawkins' failure to go through a religious experience in Persinger's laboratory suggests that the results of the latter's laboratorial techniques are basically a matter of subjective suggestibility. It is possible that this suggestibility plays an important role in the experiments.

Susan Blackmore also went through Persinger's experiment. She later reported that she did not feel anything during the first ten minutes, but that she felt under pressure to report something [Blackmore, 1994]. Eventually she felt for ten seconds an intense feeling of anger, accompanied by the sense of two hands grabbing her shoulders. Later she also felt a strong sense of anger. Blackmore's report seems to have little to do with religious or mystical experiences. It is interesting to note that at the time she went through the experiment she still was strongly involved in research on paranormal experiences and was certainly much more open to such experience than Richard Dawkins. This is also the reason why Blackmore prefers to mention such experiences more as psychical than religious.

The claim that the results of Persinger's experiments are related to the participants' suggestibility have been recently put forward by a research team at Upsala University led by Pehr Granqvist who performed the same laboratory experiments but could not find Persinger's results [Granqvist *et al.*, 2005a, 2005b]. Persinger and his research team argued that Granqvist's team did not follow exactly the same methodological procedures. For instance they claimed that the time the participants were exposed to magnetic stimulation in Granqvist's experiment was not sufficient to induce the expected experiences. This explains why they could not get the same results [Persinger, 2005]. However, Granqvist replied that his team had followed even a more precise methodology than Persinger's team. They claimed they followed a double-blind protocol: neither the participants nor the experimenters knew who was really exposed to magnetic fields. They tested two groups of volunteers: one of

43 undergraduate students, who were exposed to magnetic stimulation of the temporal lobe; and a control group of 46 equally undergraduate students, not exposed to any magnetic stimulation. Two out of three participants of the control group reported spiritual experiences. Granqvist and his team thus claimed that Persinger's experiments are so far inconclusive.

V. S. Ramachandran and Sandra Blakeslee [1998] claim that it is a well-known fact to every medical student that patients suffering from epileptic seizures in the temporal lobe, especially the left one, report intense spiritual experiences, even after the seizures stop. The authors explain that by spiritual experiences they mean direct contact with God or the felt sense of his presence, as reported by some patients. For some, it seems that a sort of a "God module" has been identified. However, both Ramachandran and Daniel Dennett are rather skeptic about the existence of such a module, since there are always several brain areas involved in every human experience [Dennett, 2006]. To presuppose that every experience related to religion is necessarily and exclusively linked to the activation of specific brain areas amounts to ignore that religious experiences are complex experiences that involve not only the brain but also, and perhaps in a more decisive way, the relations that people establish with their environments where people and culture play a crucial role. In many religions people's relation to God also plays a key role, but this relation has always something to do with other people.

Another neuroscientist, Rhawn Joseph, is basically in agreement with Persinger [Joseph, 2001]. He argues that religious behaviors, beliefs and experiences, like near-death and out-of-body experiences, trancelike states and the vision of angels and demons, are related to the activation of some specific brain areas such as the amygdale, the hippocampus and the temporal lobe.

Joseph is making here a basic mistake which, put in logical terms, consists in the fallacy of the affirmation of the consequent:

1. If someone displays a limbic system activity then he or she communed to God.
2. Jesus Christ communed to God.
3. Therefore he surely displayed a limbic system activity.

This is a well-known fallacy, since what is considered as a religious experience of communing to God may be the consequence of several different causes, including of course the activation of some brain areas. Indeed, every human experience has some relation to neural activity. However, the fact that someone reports a religious experience does not mean that such an experience is related only to the activation of some specific brain areas. It is also true that the major premise of the argument is unacceptable, since not everybody that displays a limbic system activity goes necessarily through a mystical experience. They may develop an obsessive interest in writing, drawing or in philosophizing. [Ramachandran and Blakeslee, 1998]

Andrew Newberg has avoided such a fallacy as he acknowledges: the fact that people with epilepsy in the temporal lobe reports spiritual experiences, does not allow the affirmation that such experiences are all caused by temporal lobe epilepsy [Newberg *et al.*, 2001].

One of the major issues neurotheologians have to deal with is their view of religious experience. In what follows I will be arguing that their current view is highly disputable.

### 4.1.1  *Religious Experience is Individual and Private*

The first assumption of neurotheology is an internalist and individualistic view of religious experiences that underestimates the importance and correlation that exists between personal religious experiences and external factors, namely a tradition, a community of faith sharing and the more practical aspects of most religious traditions such as the "golden rule": love your brothers as you love yourself.

Andy Clark [2000] argues that it is our interaction with culture that makes us specifically human, and he claims that we should abandon a view of human existence that is basically individualistic and brain centered. He puts forward an externalist point of view that emphasizes the importance of the environment for our understanding of how the human mind works [Clark, 2003].

Reports of religious experiences that have little or no connection at all with the real life of people sharing both the beliefs and the practices of communities are probably nothing more than the result of disturbed minds having nothing to do with religion. It is a basic mistake not to make this distinction.

Moreover, in real life, as common people grow up, they engage in love relationships; sometimes they leave the place where they may have been living for years, and move to new environments, etc., and all these new experiences may have a profound—or none!—influence upon the way people live at the religious level. They may abandon their faith; they may change from atheism to a religious belief; they may change from one religion to another, etc. Consider a person that after living for years in a rural community whose cultural beliefs and practices shaped his or her religious experiences, moves to a city where community life either disappeared or changed dramatically. This person may stop having any religious experiences like those described by Persinger, namely the sense of God's presence with no need of temporal lobe stimulation. He or she may abandon altogether any religious belief just because his/her cultural life changed significantly, but he/she may also continue to have the same religious beliefs and to pursue the same practices even if there is a change of peer group. Although this change in the person's religious experiences may have something to do with the activation of some brain areas, it also has to do, in an even stronger sense, with changes in cultural life. Persinger takes into account the cultural element only in order to explain the differences in the interpretation that people make of their laboratory religious experiences. But culture is a much richer reality that shapes human life. It includes not only beliefs and symbolic representations, but also ethical and even aesthetic practices, whereas, in Persinger's view, religious experiences seem to be a quite private issue. And he cannot avoid such a view since he reduces religious experience to the brain, and only individuals have brains.

It is true that Persinger [1987] considers that religious behaviors may not be determined by the God belief alone, but also by other factors such as the influence of peer group or of social and economic rewards. "Reward" is a key concept that according to Persinger may explain one's personal religious beliefs and practices. However, we need to take into account the possibility that people living in a given culture may develop a critical approach to religion and even overcome social influences. On the other hand, the critical attitude towards religion does not follow necessarily the reward direction. For some people, to change to a new religious tradition while continuing to live within the group that practices his or her former religion is far from being a reward.

There is in Persinger's view of religious experiences a total absence of personal, mature relations with other people. However, at least in some religious traditions the interpersonal element is a basic dimension of religious experience. Persinger agrees that there is a social dimension in religion, but he considers it only as a verbal and cultural conditioning dimension that is somehow external to the individual's religious experience itself.

Karl Peters rightly criticizes neuroscientists such as Eugene d'Aquili and Andrew Newberg who in their book *The Mystical Mind* investigate the neurobiological bases of religion, precisely because they study religious experiences that are all individual experiences, which seem to have little to do with the ordinary daily life of common people [Peters, 2001].

### 4.1.2 *Religious Experiences Are Basically Connected to Out-of-This-World Entities*

According to Persinger [1987], during religious experiences people usually feel that they come into contact with some out-of-this world entity, such as a cosmic mind, a god, some extraterrestrial being, etc. Thus religious experience typically involves strange, bizarre and unusual contacts with such entities. The typical religious experience is the sensed presence of some non-physical entity, something that was reported by the participants having their temporal lobe artificially activated during his laboratory experiments [Persinger, 2003]. But this is a rather arbitrary and indeed simplistic view of religious experience. To sense the presence of someone that is identified with God is not necessarily a common and prototypical religious experience. If one keeps his or her faith despite feeling a deep and radical absence of God, such an attitude may count as a deep religious experience, and it would be interesting to know which brain area is responsible for the "sense of absence."

Moreover, the author clearly identifies altered mental states with religious experiences. The experience of "sensed presence" is associated by Persinger with several types of altered mental states which he also wrongly identifies with religious experiences. This is clearly seen in the description that a 30 years old woman, whose brain was put under the effect of a magnetic field, makes of her altered mental states:

> I feel detached from my body... I am floating...there is a kind of vibration moving through my sternum... there are odd lights or faces along my left side. My body is becoming very hot... tingling sensations in my chest and stomach... now both arms. There is something feeling my ovaries. I can feel my left foot jerk. I feel there is someone in the room behind me. The vibrations are very strong now and I can look down and see myself. [Persinger, 2003: 281]

We find in this woman's experience nothing but altered mental states, which have very little, if anything, to do with truly religious experiences.

Ramachandran reports also the case of Paul, a patient with epileptic seizures that came to see him:

> There is a soft armchair in our laboratory, but Paul seemed unwilling to relax. Many patients I interview are initially uneasy, but Paul was not nervous in that sense—rather, he seemed to see himself as an expert witness called to offer testimony about himself and his relationship with God. He was intense and self-absorbed and had the arrogance of a believer but none of the humility of the deeply religious. With very little prompting, he launched into his tale. "I had my first seizure when I was eight years old," he began. "I remember seeing a bright light before I fell on the ground and wondering where it came from." A few years later, he had several additional seizures that transformed his whole life. "Suddenly, it was all crystal clear to me, doctor," he continued. "There was no longer any doubt anymore." He experienced a rapture beside which everything else paled. In the rapture was clarity, an apprehension of the divine—no categories, no boundaries, just an Oneness with the Creator." [Ramachandran and Blakeslee, 1998]

Because both Ramachandran and Persinger mention only religious experiences of people with clear mental disorders, they can only make an unwarranted generalization about religious experiences of people who are not mentally disturbed. A clear distinction needs to be made between normal and pathological spiritual experiences. We also need to acknowledge that religious behaviors and beliefs have to be necessarily mystical. [Newberg *et al.*, 2001]

### 4.1.3  *Religious Experiences Are Basically Emotional and Positive*

According to Persinger [1987] a religious experience is typically characterized by positive emotions such as euphoric feelings, peace and reduction of death anxiety, although he believes that negative emotions, like fear of death, may also be involved. He does not consider that cognitive elements are constitutive of religious experiences, and never mentions any critical and conscious dimension of religion. It seems that people are always under both emotional and cultural pressure, and that this pressure does not allow any critical discernment. But it is clear that the cognitive and critical dimensions of religion are often present. The prophets of the Old Testament in the Judeo tradition represent this critical instance that is constitutive of the Jewish tradition.

Persinger persistently ignores that there is indeed critical thinking in at least some of the major world religions, and that there have been indeed, as in the case of the Bible, significant changes in its interpretation on the basis of scientific progress. The author chooses either to ignore or to give little importance to this fact because that would of course weaken his whole argument.

Neurotheologians tend to overemphasize the emotional aspect of religious experience. This is however an oversimplification. People may and indeed need to reflect about their religious beliefs and about their religious behavior and experiences.

On this issue, N. Azari and her research team found that religious behavior are correlated with brain areas that play an important role in cognition [Azari *et al.*, 2001]. Using functional neuroimaging Azari's group has also found that religious experience is cognitively mediated and that there are some cognitive elements in religious experience, since brain areas activated correspond to neural areas involved in both emotion and thought. Azari has explicitly criticized some authors such as Joseph and Persinger because of their views on the emotional nature of religious experiences [Azari *et al.*, 2005]. She believes that the results of her research team point in a quite different direction. Religious experiences are not only emotionally driven experiences. Their cognitive dimension makes it possible that a critical element may free religious traditions from beliefs and practices that are unnecessary. The critical element of religion also allows us to distinguish between pathological and non-pathological religious beliefs and behaviors.

Moreover, the emotional component of religious experience may not always be pleasant and positive. Finding God during frustrating experiences, for example, may be a very deep religious experience [Woodward, 2001].

### 4.1.4 *Causation and Correlation*

Persinger seems to identify correlation with causation, and to attribute to both the same ontological and epistemological values. Already in his book *Neuropsychological Basis of Human Belief* (1987) Persinger [2003] mentions that god's beliefs are *determined* by psychological and neurobiological factors. More recently, he claimed that not only religious but every kind of human experience is *generated by determined* and *correlated with* neural activity. He identifies *correlation* with both *causation* and *determination*.

However, most stimuli that activate the brain come from the environment, and the different stimuli evoke different brain reactions. Thus, many, if not most, of our experiences are the result of *a correlation* between environmental stimuli *and* brain activity as a reaction to such stimuli. Thus, one may not identify neural correlation with neural causation. Persinger, however, although recognizing that there are environmental stimuli involved in every human experience, reduces religious experience to mental events and the cause of such events to brain activity. I have already pointed out: that there is a neural correlate of a given behavior does not mean that such a correlate is *the* (only) cause of such a behavior. In most cases, human behavior is the result of a complex set of elements that include brain activation as a response to external stimuli, and also brain activation that not only produces emotions but also allows rational argumentation and freedom of decision, as responses to such external stimuli. Correlation does not necessarily mean causation, since a given correlation may not produce, by itself alone, a specific effect, such as a religious experience.

Religion has several complex dimensions that have to do not only with neural correlates but also with social and cultural life, with life of prayer, works of charity, the promotion of peace through education, dialogue and tolerance, etc., that are important non-neural causes. There are no simple neural-causal determinations of religion. There is room for reflection, dialogue and free decisions. Thus, brain areas associated by neurotheologians with religious experiences, such as the temporal lobe, do not always *determine* or *cause* such experiences. On the other hand, the felt absence of God, doubts about God's existence, acts of love towards the others, are all religious experiences that are not necessary correlated with the temporal lobe.

Finally, there is no reason to deny the possibility that there is in truly religious experiences an interaction with a divine entity generally called god. Ramachandran mentions this hypothesis although he considers that it is not an interesting one, since it neither can be proved nor disproved on the basis of any empirical evidence [Ramachandran & Blakeslee, 1998]. But what would count as an empirical evidence of God's interaction with human beings? If God is not one among the many elements in our universe, he surely cannot be detected by empirical tests. Should there be any of such tests, then God wouldn't be God, since he would be one among the many elements of our universe. Thus, the hypothesis that there is indeed a divine entity cannot be ruled out [Albright, 2003].

## 4.2  Self, Soul and Human Immortality

In his work *The Astonishing Hypothesis*, Francis Crick argued that a human being is nothing but a bundle of cells, and therefore the concept of a soul has been abandoned even by many Christians. Antonio Damasio argues that research in neuroscience undermines the traditional concept of a substantial and permanent self. The self is a mental creation and it is a neurological process [Damasio, 2003], a view that is in agreement with the Buddhist tradition. Gerald Edelman has the same view and draws the conclusion that since there is no metaphysical and enduring self, there can be no survival after death. Christopher Koch however believes that, the fact that there is neither a self nor a soul allows human beings to hope for immortality [Koch, 2004].

For centuries, at least in the West, many people believed, and still believe that human beings are a unity of body and soul. Although united, they are separable at the moment of death. Pope John Paul II confirmed in the nineties that the Catholic Church believes that the soul is created by God at the moment of the biological conception of a new human being. How are we to reconcile the view that there is no metaphysical or spiritual soul with the Christian belief that there is life after death? If with death human beings disappear completely, what is the point of believing in God?

## 4.3 Theological Consequences

Andrew Newberg argues that neurotheology represents no threat to traditional religions [Newberg *et at.*, 2001]. On the contrary, he says, neuroscience will allow a better understanding of religion and indeed will help to reconcile science and religion. Newberg seems to acknowledge that religion has to do with more than what goes on inside the human brain. It has to do with community and family life, ethics, love, compassion and forgiveness. However, he also claims that the role of neurotheology is to allow us to understand the connection between brain activity and *all* (Newber's italics) main elements of religion. However, the author is relying on the work that has been done so far by neuroscientists and himself on the basis of both medical practice, such as that of Ramachandran, and laboratorial experiments such as those of Persinger and Joseph. But such experiments are based on highly disputable views on what counts as a religious experience as I have argued before. To repeat, Ramachandran's examples are those of his mentally disturbed patients whose "mystical" experiences cannot be identified with religious experiences of common people. The experiences reported by such patients are typically bizarre; they are related to out-of-this-world entities and are also self-centered. Such experiences fit well in the internalist view of the human person supported by both neuroscientists and neurotheologians. This view contradicts most world religions, and is indeed a caricature of religion.

There is however an acceptable theological consequence of neuroscience. It helps religions to acknowledge that the old traditional dualisms such as body and soul, material/spiritual, natural/supernatural, immanent/transcendent, life before death/life after death, etc., need to be re-evaluated. Such dualisms have led many religious people to look at "this world" with caution and even with fear. Religious experience may indeed appear more linked with another, supernatural world than with "this word." The fact that religious experience takes place not only in this world but also "in the flesh," helps to unify human life and to avoid ontological dualisms that make it difficult to understand how God is related to humanity. An embodied religion is probably the more adequate way to understand religion in the 21$^{st}$ century. This is a new paradigm and a new challenge that traditional theologies have yet to face seriously.

Nancy Murphy [1998], a Christian theologian, has proposed a non-reductive physicalism—the view that the physical dimension of human beings is all they need to enter into relationship with God. Joseph Ratzinger [2004], the actual Pope, has already put forward an alternative view to the more traditional one of the body-soul dualism. He believes it is possible to abandon such a substantial dualism in favor of a relational view of the soul and of immortality. He also feels no need of the natural/supernatural dualism.

Finally, Andrew Newberg has put forward the hypothesis that the human brain has evolved in such a way that it developed neural structures that allows humankind to conceive of transcendental realities. While this hypothesis is surely rather unexpected, it has no special consequences for those religions that admit the existence of a transcendent God, wherein, the immanence/transcendence dualism is not ontologically needed.

# References

Albright C. R. & Ashbrook, J. B. [2001] *Where God Lives in the Human Brain* (Sourcebooks, Naperville, IL).

Albright, C. R. [2003] "Religious experience, complexification, and the image of God", in *Neurotheology: Brain, Science, Spirituality, Religious Experience*, ed. Joseph, R. (California University Press, Berkeley, CA) pp. 167-182.

Alper, M. [1996] *The "God" Part of the Brain* (Rogue, New York).

Ashbrook, J. B. & Albright, C. R.. [1997] *The Humanizing Brain: Where Religion and Neuroscience Meet* (Pilgrim, Cleveland, OH).

Azari, N.P. *et al.* [2001] "Neural correlates of religious experience," *European Journal of Neuroscience* **13** (8), 1649-1652.

Azari, N. P. *et al.* [2005] "Religious experience and emotion: evidence for distinctive cognitive neural patterns," *The Int J Psychol Relig* **15** (4) 263-281.

Beauregard, M. and Paquette, V. [2006] "Neural correlates of a mystical experience in Carmelite nuns," *Neurosci. Lett.*, **405** (3) 186-190

Blackmore, S. [1994] "Alien abduction: The inside story," *New Scientist*, November 19, **1952**, 29-31.

Clark, A. [2000] "Making moral space: A reply to Churchland," in *Moral Epistemology Naturalized*, ed. Campbell, R. & Hunter, B. (University of Calgary Press, Calgary) pp. 307-312.

Clark, A. [2003] *Natural-Born Cyborgs* (Oxford University Press, Oxford).

Damasio, A. [2003] *Looking for Spinoza: Joy, Sorrow and the Feeling Brain* (Harcourt, Orlando).

D' Aquili, E. & Newberg, A. B. [1999] *The Mystical Mind: Probing the Biology of Religious Experience* (Fortress, Minneapolis).

Dennett, D. [2006] *Breaking the Spell: Religion as a Natural Phenomenon* (Viking Penguin, London)

Ganqvist, P. *et al.* [2005a] "Sensed presence and mystical experiences are predicted by suggestibility, not by the application of transcranial weak complex magnetic fields" *Neurosci. Lett.* **379** (1) 1-6.

Ganqvist, P. *et al.* [2005b] "Reply to M.A. Persinger and S. A. Koren's response to Granqvist et al. 'Sensed presence and mystical experiences are predicted by suggestibility, not by the application of transcranial weak complex magnetic fields' " *Neurosci. Lett.* **380** (3) 348-350.

Joseph, R. [2000] *The Transmitter to God. The Limbic System, The Soul, and Spirituality* (California University Press, Berkeley, CA).

Joseph, R. [2001] "The limbic system and the soul: Evolution and the neuroanatomy of religious experience," *Zygon* **36**, 105-136.

Koch, C. [2004] *The Quest for Consciousness: A Neurobiological Approach* (Roberts, Englewood, CO).

88     *A. Dinis*

Murphy, N. [1998] "Nonreductive physicalism: Philosophical issues," in *Whatever Happened to the Soul? Scientific and Theological Portraits of Human Nature*, ed. Brown, W., Murphy, N. & Malony, H. (Fortress, Minneapolis) pp. 127-148.

Newberg, A., d'Aquili, E. & Rause, V. [2001] *Why God Won't Go Away* (Ballantine, New York)

Persinger, M. [1987] *Neuropsychological Basis of Human Belief* (Praeger, New York).

Persinger, M. [2003] "Experimental simulation of the God experience: Implications for religious beliefs and the future of the human species," in *Neurotheology: Brain, Science, Spirituality, Religious Experience*, ed. Joseph, R. (California University Press, Berkeley, CA) pp. 279-292.

Persinger, M. *et al.* [2005] "A response to Granqvist *et al.*'s 'Sensed presence and mystical experiences are predicted by suggestibility, not by the application of transcranial weak magnetic fields'," *Neurosci. Lett.* **380**(3), 346-350.

Peters, K. [2001] "Neurotheology and evolutionary theology: Reflections on the mystical mind," *Zygon* **36,** 493-500.

Ramachandran, V.S. & Blakeslee, S. [1998] *Phantoms in the Brain* (Quill William Morrow, New York).

Ratzinger, J. [2004] *Introduction to Christianity* (Ignatius, San Francisco).

Woodward, K. [2001] "Faith is more than a feeling," *Newsweek*, May 7, 58.

# 5

# SciComm, PopSci and The Real World

*Lui Lam*

A physicist's experience in science communication (SciComm), popular science (PopSci) and the teaching of a Science Matters (SciMat) course *The Real World* is presented and discussed. Recommendations for others are provided.

## 5.1 Introduction

Yes, yes, I know. I know that I am not supposed to use abbreviations in a chapter title; I should spell out the whole word. But like the French say: rules are set to be broken. And indeed it happened: Newton (1643-1727) broke the rules set by Aristotle (384-322 BC) in dynamics, and replaced them with his own three laws; Einstein (1879-1955) in turn broke Newton's three laws and replaced them with his theory of special relativity. This is called innovation or in rare occasions, revolution. Rules could and should be broken when one has a good reason. And I have *two* good reasons.

My background as a scientist is not atypical. I have been working in physics research in the last 40 years. I am now a professor in California, a job involving both research and physics teaching (with an unbelievable teaching load of 12 credits[1] plus office hours per semester). My research was first in condensed matter physics and later in nonlinear physics and complex systems.

---

[1] At San Jose State University, an undergrad lab of 3 hours is counted as 2 credits (*versus* 3 credits in the community college of City University of New York, a great city) as the instructor's teaching load is concerned. I end up teaching 2 courses and 3 labs per week.

My involvement in Science Communication (*SciComm* or *scicomm*) began in 1994, in Mexico City, Mexico. In that year I was invited by Rosalio Rodriquez and gave a public lecture "Nonlinear Physics Is for Everybody" at Universidad Nacional Autonoma de Mexico. Since then, I have been doing physics research, teaching and scicomm simultaneously, trying to synthesize my activities in these three areas and, most importantly, trying to be creative and have fun in doing that. In recent years, these activities are heavily influenced by my involvement in histophysics—the physics of human history [Lam, 2002; 2008b] and Science Matters[2] (SciMat or scimat) [Lam, 2008a]. What follows is the adventure I went through in the wonderland of SciComm and my innovation in education on behalf of SciMat, and my recommendation for others.

## 5.2  Science Communication

Science communication [Gregory & Miller, 2000] involves four components:

1.  Funding and organized effort from the government and learning societies
2.  Engagement of scientists as individuals
3.  Participation of the public
4.  Development of SciComm as a research discipline, by scholars and students

Engaging scientists and the science community to participate actively and regularly is a daunting task. What the government can do is to provide funding and encouragement to scientists who are willing and qualified. The other part of the game concerns the scientists themselves, at the individual level. Leon Lederman, a Nobel physicist, proposed that working scientists should devote 10% of their time to communicating science. This may not be very practical for those professors who do not yet have tenure, because the competition in research is very keen and

---

[2] Science Mattes is a new discipline that treats human-related matters as part of science.

research requires undivided attention, not to mention that SciComm is not always appreciated and rewarded by the administrators. But let us say, a scientist—tenured or not—wants to contribute to SciComm, what can she or he do? This Section addresses this problem, from the perspective of a working physicist.

Six items concerning what science professors or teachers can do in SciComm are presented here [Lam, 2006a].

1.  <u>What every science professor/teacher can do: Integrate popular science books into science teaching</u>

The quick pace of interdisciplinary development in science and the ever-changing job market demand a broad knowledge base from our students. For five or more years, I integrated popular science (*PopSci* or *popsci*) [3] books[4] into my physics classes by giving extra credits to the students who would buy a popsci book,[5] read it and write up a report [Lam, 2000a; 2001; 2005a]. The instructor does not actually teach the books, and hence will *not* find the teaching load increased—an important factor in any successful education reform. It is like a supplementary reading, a practice commonly found in English classes but rarely adopted by science instructors. The aim of this practice [Lam, 2000a] is to:

(1)  Broaden the knowledge base of students
(2)  Show the students the availability and varieties of popsci books in their local book stores
(3)  Encourage the students to go on buying and reading at least one popsci book per year for the rest of their life
(4)  Become a science-informed citizen—a voter or perhaps a future science-friendly legislator

It is about lifetime learning of science matters. Professors in other universities have copied this approach, with equal success. It is equally applicable to high schools. Adopting this practice in the whole country or worldwide in large scale will fundamentally improve the science

---

[3] The term PopSci is inspired by Pop Art, advocated by Andy Warhol (1928-1987).
[4] See Section 5.4.1 for the reasons of why popsci books are important.
[5] See Appendix 5.1 for a sample of books bought by my students.

education of our students, the future average citizens. An immediate side effect is that in a few short months, all the popsci books on the bookshelves of every bookstore will be wiped out. The popsci book market will be drastically improved, attracting more skillful writers into the popsci books profession, benefiting everybody.

2.  <u>What every science professor can do (I): Inject popular science talks into departmental seminars, or set up a separate popular science seminar series in the department</u>

Since 1994, I have been giving public talks on science, history and religion, starting with a title the audience are interest in and leading them to the topics such as the scientific method that I really want them to learn. The titles include:

- Wu Chien-Shiung: The First Woman President of American Physical Society
- Does God Exist?
- The Real World
- The Birth of a Physics Project: What Happened to My New Book
- Why the World Is So Complex
- How to Model History and Predict the Future

I usually tried them out first in my physics department. In almost all universities around the world, there is a weekly departmental seminar. Recent research results are presented by either outside speakers or the faculty members. These talks are usually boring and quite often poorly attended. The exceptions are popsci talks, because they are easy to understand, even for undergraduates.

What every science professor can do is to insert popsci talks into their departmental seminar series, which can be given by themselves or outsiders. If the department does not allow it, a separate popsci seminar series can be set up within the department, with the help of the student science club if it exists—if not, help the student to set up such a club; your department chair will be thankful. And, of course, these seminars are open to the general public.

3. What every science professor can do (II): Set up a popular science lecture series in the university for general audience

In December 1999, I established a public lecture series "God, Science, Scientists" at San Jose State University (SJSU). The first three speakers (Fig. 5.1) are:

(1)  Michael Shermer, who gave a talk in May 2000 on "How People Believe: The Search for God in the Age of Science." Shermer, a monthly columnist for *Scientific American*, is the founding publisher and editor of *Skeptic* magazine. He is the author of many popsci books such as *Why People Believe Weird Things*, *How People Believe, Denying History, The Borderlands of Science, The Science of Good and Evil, Why Darwin Matters* and *The Mind of the Market*. He is also a professor of history and science associated with Caltech and the Occidental College at Los Angeles.

(2)  Eugenie Scott, the executive director of the National Center for Science Education in El Cerrito, California. Scott is a nationally known authority on creationism and evolution controversy.

(3)  Charles Townes, the Nobel laureate in physics and co-inventor of laser.

These talks were attended by a large audience from different walks of life and were well received. I still get letters/emails from the fans who attended the lectures.

Every science professor can set up a popsci lecture series in their university, which will be highly appreciated by the administrators. It is not that difficult to do if you limit yourself to one speaker per semester. And don't forget to invite your Dean or President to introduce the distinguish speakers.

Fig. 5.1. The first three speakers of the "God, Science, Scientists" public lecture series at SJSU. From left to right: Michael Shermer, Eugenie Scott and Charles Townes.

### 4. What every scientist can do: Give popular science talks in high schools, the community and other places

For a period of 11 years, I gave invited popsci talks in various high schools,[6] universities, TV interviews (CCTV, Dec. 18-19, 2003) and conferences in Mexico, the United States, Taiwan, Hong Kong and China.

In November 2000, Shermer [2001] was one of four PopSci experts I invited, in my capacity as a member of the International Advisory Committee, as a speaker at the International Forum on Public Understanding of Science, Beijing, organized by China Association for Science and Technology (CAST). We became good friends. I wrote an article on active walk for his magazine *Skeptic* [Lam, 2000b].

This article led to an unexpected invitation from the Foundation For the Future (in Bellevue, WA), as a keynote speaker in their annual seminar, Humanity 3000, held in Seattle, 2001. The 23 invited "participants" included the famous Edward O. Wilson (from Harvard) and Richard Dawkins (from Oxford); I was the only physicist there. I gave a talk on "How to Model History and Predict the Future" [Lam, 2003], and became a futures-study expert, *ipso facto*.

---

[6] Such as the Provincial Senior High School, Hsinchu, Taiwan, whose graduates include Yuan-Tseh Lee (Li Yuanjie), Nobel laureate in chemistry.

After that, I was invited by Doug Vakoch of the SETI Institute (Search for Extraterrestrial Intelligence, based in Mountain View, CA), who also attended this Seattle seminar as an "observer," to go to Paris in March 2002 and talk about what science-and-art message to send to the extra-terrestrials (ET), in case they exist. I proposed to beam them digitally the recipes to create the Sierpinski gasket, a fractal.[7] Vakoch liked the idea and included it in his workshop report [Lam, 2004a]. And I suddenly found myself an ET expert.

One thing led to another, like in a chain reaction. I met some artists during this Paris workshop, and we have been trying to collaborate on a physics-art-music (PAM) project[8] called "Candle in the Wind."

Another participant in that Seattle seminar was Clement Chang, founder of Tamkang University in Taiwan. In December 9-11, 2003, I was invited to give the Tamkang Chair Lectures (Figs. 5.2). My host was Kuo-Hua Chen, Chair of the Graduate Institute of Futures Studies and Director of the Center for Futures Studies. The result is my first popsci book, *This Pale Blue Dot: Science, History, God* [Lam, 2004b] (Fig. 5.3).[9]

It is not true that every science professor is good at giving popsci talks, but every one can try and be successful. You just keep practicing, giving the same talk many times and modifying it with the help of PowerPoint. And as shown in my story above, the reward could be significantly large: It gains you many new friends, from all walks of life; it might even take you to Paris.

---

[7] A fractal is a self-similar mathematical or real object with possibly a fractional dimension [Lam, 1998].

[8] The artist, Aprille Glover (www.aprille.net), and her husband are two Americans living in Lavardin, France [Glover, 2000].

[9] "Pale Blue Dot" refers to our dear Earth when observed from far, far away in space; it comes from the title of Carl Sagan's popsci book [Sagan, 1994]. My book contains three chapters: Why the World Is So Complex, How to Model History and Predict the Future, and Does God Exist?

Fig. 5.2. The poster of my Tamkang Chair Lectures, titled "This Pale Blue Dot: Science, History, God."

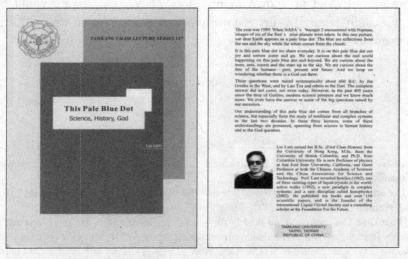

Fig. 5.3. The covers of my first popsci book *This Pale Blue Dot: Science, History, God.*

## 5. What some scientists can do but all can try: Contribute to science communication research

Science communication as a discipline is at its very early stage; it is a profession without a formal name[10]—unlike the case in physics, say. A new and short word is needed. My suggestion is to call it SciComm or PopSci.

It is rare to find a scicomm course in American universities. In contrast, China has a lead here; there are already degree programs in SciComm in at least four universities, and a research institute on PopSci (under CAST) in Beijing—the China Research Institute for Science Popularization. Obviously, the contribution of working scientists in making PopSci a mature discipline is much needed; for example, they can provide different perspectives and help to clarify science issues.

In June 2004, I collaborated with Da-Guang Li of CAST (now at Chinese Academy of Sciences, or CAS for short) and Xu-Jie Yang of the *ScienceTimes* (a Beijing daily published by CAS) and presented a paper at the International Conference on Scientific Knowledge and Cultural Diversity, Barcelona, Spain, June 3-6, 2004, on the absence of *professional* popsci book authors in China [Lam *et al.*, 2005] (see Section 5.4). This was followed by a paper on a new concept for science and technology museums, presented at the International Forum on Scientific Literacy, Beijing, July 29-30, 2004 [Lam, 2006b]. The idea is that unified themes governing natural and social sciences [Lam, 2008a] should and can be injected into the display in science museums, to avoid the possible misconceptions conveyed to the visitors that the two are completely separated from each other (see Section 5.3). And, reporting for *ScinceTimes*, Yang and I co-wrote an article reporting on the 10th International Conference on the History of Science in China, Harbin, August 4-7, 2004 [Yang & Lam, 2004].

---

[10] The absence of a formal name for the SciComm discipline or profession is due to the fact that the practitioners cannot agree on a single name, partly due to the shifting emphasis or concept in SciComm. Some favor Popular Science or Science Popularization; others, Public Understanding of Science; etc. In fact, these different terms could be the names of subfields within a single discipline—SciComm, like atomic physics and condensed matter physics, two subfields in physics.

6. What some science professors can do: Merging science with humanities

Science and the humanities are considered by some as "two cultures" [Snow & Collini, 1998; Lam, 2008a]. But in fact, humanities are about humans, which is a (biological) material system of *Homo sapiens*. Thus, humanities could and should be part of the natural sciences, which is about *all* material systems. The two can be integrated, but how?

In 1992, two years after I founded the International Liquid Crystal Society [Lam, 2005b; 2005c], I came up with a new paradigm for complex systems. I named it *active walks* (AW), reviewed in [Lam, 2005b; 2006c]. An active walker is one that changes a landscape—real or mathematical—as it walks; its next step is in turn influenced by the deformed landscape. Active walk is now widely applied in natural and social sciences ([Lam, 2008a; Han *et al.*, 2008]).

By 2000, the year that Shermer and I first met each other, I have been trying to create a new discipline by merging AW with a branch of the social sciences/humanities. Contact with Shermer, himself a historian [Shermer & Grobman, 2000], made me look at history seriously. Two years later, I presented my *first* paper [Lam, 2002] on the physics of history, or *histophysics* [Lam, 2008b], at the workshop celebrating the 80th birthday of Chen Ning Yang, a physics Nobel laureate, at Tsinghua University, Beijing. Histophysics is a successful example of SciMat, the new discipline that treats all human-related matters as part of science [Lam, 2008a]. My work in histophysics leads us to the discovery of two historical laws concerning Chinese dynasties (from Qin to Qing) and a new general phenomenon in Nature called the *bilinear effect* [Lam, 2006c; 2008b; Lam *et al.*, 2008]. My knowing of Shermer, made possible through our shared activities in SciComm, played an important role in the creation of this new discipline, histophysics [Lam, 2005d]. Subsequently I expanded my interest from history to the overall basic situation of humanities/social science and came up with the idea of SciMat [Lam, 2008a; 2008c].

In the summer of 2005, I presented a paper on the history of histophysics [Lam, 2005d] and worked as a reporter for *ScienceTimes* [Lam, 2005e; 2005f] at the XXII International Congress of History of

Science, Beijing, China, July 24-30. I met Maria Burguete, another participant from Portugal. Upon a cup of Chinese tea, she invited me to visit her. Next year in March, with the award of US$ 1,000 travel money from SJSU, I found myself in Portugal, one of the few countries in Europe that I never visited before, despite my two-year stay in Belgium and West Germany during 1975-1977. It was at the bar of Vila Galé in Ericeira and after a few drinks that we decided to do something together next year, and that was how the First International Conference on Science Matters, Ericeira, Portugal, May 28-30, 2007, co-chaired by Maria Burguete and Lui Lam, came about [Sanitt, 2007].

My involvements in SciComm actually include something more. To help China's fight against pseudoscience, and sometimes "evil religions," I became the Chinese-copyright agent of Michael Shermer and James Randy. I got Shermer's *Why People Believe Weird Things* (Hunan Education Press, 2001) and Randy's five books on magic and pseudoscience fighting (Hainan Press, 2001) published in Chinese.

There are many things scientists can do in SciComm, as individuals and without funding. Six of these are recommended above, with the first four suitable even for untenured professors. SciComm is fun and adventurous; it enables one to meet interesting new friends/colleagues beyond their own discipline, or even helps one's research career. Chair Mao once said: When faced with a daunting task, learn from the ants; mobilize the masses and trust them. It worked for China, and will work for SciComm.

## 5.3 A New Concept for Science Museums

A science museum (or a science and technology museum) is an effective medium in helping the public to understand science. However, in contrast to popsci books [Lam, 2001; 2005a; Lam *et al.*, 2005] and TV science programs, museums are limited by their physical locations and large budgets. Yet, when available these museums allow the public to see the real objects and, apart from admiring the wonders of Nature itself, learn the science principles behind some natural phenomena.

In China new science museums appeared rapidly in the last 20 years. In other parts of the world, for example, in Barcelona, Spain, a brand new science museum is under construction. There is no doubt that the importance of science museums is well recognized.

The first step in making a good science museum is to have good exhibits. The next step is to make it physically interactive, partially or completely. Almost all science museums *stop* here. This could create a problem and is most unfortunate; most unfortunate because the problem is easily removable. What is needed is a new concept.

### 5.3.1 *Possible Misconceptions Imparted to the Visitors*

The exhibits in all science museums are displayed according to their subject matter, in other words, in compartments. For example, the exhibits may be put into four divisions: inanimate matters, life, intelligent matters, and civilizations. This classification is based upon the hierarchic construction of the material world, according to what we know. The world is made of atoms; in increasing size, atoms form molecules, molecules form condensed matter—inorganic matters and organic matters. Organic matters form living matters—plants and animals. Animals consist of cells and organs. In particular, we have human bodies. A group of humans form a society, leading to civilizations. (See Fig. 1.2 in [Lam, 2008a].) Consequently, the four divisions of the exhibits are logical and there is nothing wrong with that. However, science museums with these compartmental exhibits could create two misconceptions for the visitors:

1. The visitor may leave with the impression that science is neatly divided into compartments; that is, there is no unifying themes or principles behind many of those exhibits.
2. Since almost all science museums are limited to natural sciences only, the visitor may go home thinking that there is a rigid demarcation separating the natural sciences from the social sciences.

The fact that social science should and could only be based on natural science [Lam, 2002; 2006c; Wilson, 1998] is easy to see, but is sometimes overlooked. As explained in last Section, the reasoning goes like this: Social science is about the study of human behaviors and human societies. Humans are (biological) material bodies which, of course, are part of natural science since natural science is about *all* material systems. (See [Lam, 2008a] for more discussion.)

### 5.3.2 *A Simple Remedy*

How can these two misconceptions be avoided and corrected? Very simple! Before the exit of every science museum, there should be a room or a space showing some established principles that are able to unify many different phenomena found in Nature, with examples taken from both natural and social sciences. There are three such principles: fractals, chaos and active walks [Lam, 1998]. (See [Lam, 2008a] for a brief introduction to these three general principles.)

It is gratifying to note that in some science museums in China[11] [Ai, 2004][12] and perhaps elsewhere, some, but not all, of the three general principles mentioned above have been included in their exhibits. However, there is still no emphasize on the theme that social science and natural science are an integral whole, and the former is based on the latter, with unifying principles. And we would like to see that this is the case in *all* museums in the world.

Lastly, to have the greatest and lasting impact on the visitors, I still think that putting the unifying themes concerning all natural and social phenomena before the exit of a science museum is the best choice.

### 5.4 Science Popularization in China

In China, the term "popular science" or "science popularization" (abbreviated as *kepu* in Chinese) is favored over "science communication," due mostly to the fact that the former two terms (especially the second one) have been in use for a long period of time. A

---

[11] China Science and Technology Museum, Beijing: "Science Tunnel" (http://old.shkp.org.cn/xinxi/suidao/shuidao003.htm).

[12] Ai's article is an introduction to the Shandong Science and Technology Museum in Jinan, Shandong Province, China.

brief history of science popularization in China, from the time of late Qing Dynasty and up to 2006, can be found in [Li, 2008].

Essentially, before 1949, the year the People's Republic of China was established, PopSci was advanced by the intellectuals with hands free from the government; many of these people were educated in the West or Japan. After 1949, like everything else in the New China, PopSci was managed from the top by the government. The advantage is that PopSci is financially secure; the disadvantage, as pointed out by Li [2008], is that there were less free discussion and exchange of idea among the practitioners or scholars. As mentioned in item 5 of Section 5.2, in SciComm, China actually has a lead over many other countries in terms of scales. A summary of the current situation—official policies, programs, activities and studies of PopSci in China is available.[13] Those interested in PopSci research in China could consult the journal *Science Popularization*[14] which is based in Beijing.

Here is an interesting PopSci problem: Why *professional* popular-science book authors do not exist in China? The easy answer to this question would be that, like some other non-English-speaking countries, the sale of popsci books written not in English (and hence no worldwide sales) is not enough to support their authors full time. But China is a huge country with 1.3 billion people. The story is more complicated than this. The answer to and solution of the problem in China's case could be unique.

Before we proceed to the answer, let us first review why this question is important, not merely to China but to the whole world. And, after the answer, recommendations to improve the situation, applicable to China and *beyond*, will be given.

### 5.4.1  *The Importance of Popular-Science Books*

Popular science books have a long history in existence [Gregory & Miller, 2000]. Unfortunately, they are a neglected tool in the science

---

[13] *2007 Science Popularization Report of China*, published by China Research Institute for Science Popularization, CAST (Popular Science Press, Beijing).

[14] This journal is managed by China Research Institute for Science Popularization, CAST. Since its inception in 2006, the author is a member of the editorial board.

education of students and ordinary citizens [Lam, 2005a]. Popsci books are unique among the science media:

1. They are available in every bookstore in every town, unlike the technical science books which are available in special book stores in a university town.
2. Many popsci books are written by the pioneers themselves, Nobel laureates, or very gifted science writers who could be journalists or other scientists.
3. These books are affordable to almost everybody (about 20 yuans in China, and 15 dollars for a paperback in USA).
4. These books are the place to learn how research was actually done and how discoveries were made in very recent times.
5. These books, at least in the USA and for the majority of them, contain no equations; they, if well written, are easy and entertaining to read.

Obviously, to ensure the continuous supply of new and good popsci books, a large number of competent authors must be available.

### 5.4.2 *Popular-Science Book Authors in China*

In spite of China's large population of 1.3 billion, there is not yet a single *full-time* professional popsci book author in this vast country. This is in contrast to the case in literature, because China does have professional writers who can support themselves by publishing novels. And this is not due to lack of support from the Chinese government. In fact, the Chinese government recognizes science and technology as an important pillar in raising the living standard of its population and the economic well-being of the country as a whole. In 2002, China passed the *law*,[15] the one and only one such law in the world, which protected and encouraged science popularization at every level of government.

---

[15] *Law of the People's Republic of China on Popularization of Science and Technology*, issued June 29, 2002 (Popular Science Press, Beijing).

In the years from 1949 to about 25 years ago and *before* market economy was introduced, every writer in China was government employed. During this period of time, the government saw the need to support full-time novelists, but not full-time popsci writers. Obviously in China (and everywhere else in the world) popsci books are not deemed to be equally important as literary books.

These days, *after* market economy is in place, quite a number of self-employed literary writers already exist and, as usual, *the government still supports a sizable number of literary writers*. Yet, we still see no full-time popsci book authors in China, self-employed or government employed. Why? To find out what happened, we interviewed a number of popsci book authors and publishers in China [Lam *et al.*, 2005]. We were told that:

1.  Science popularization is considered lower in status compared to science research or teaching.
2.  Work in science popularization is not counted in job evaluations in many places.
3.  Lack of systematic and large-scale government effort or program to train popsci professionals.
4.  Insufficient personal income to support free-lance, full-time popsci writers.

Points 1 and 2 are definitely true in almost every other country; some countries are doing something to tackle point 3; point 4 is untrue, for example, in USA.

Point 4 is particularly interesting. With such a huge population in China, how can this happen? In fact, presently, the sale of an average popsci book in China is less than 5,000 copies. There are exceptions: for example, *The Complete Book of Raising Pigs* did sell 3 million copies. What this implies is that a popsci book (not on pig raising) geared to the need of the masses is still waiting to be written.

### 5.4.3 *Recommendations*

To address points 3 and 4 above, here are six recommendations:

1. The government should recognize the importance of popsci books, in line with the popsci law they put into effect in 2002, and support popsci writers the same way they support literary writers.
2. The government could extend the policy of supporting literary book projects to popsci books, too. That is, prospective writers can apply for a grant to write a particular popsci book.
3. In every science funding agency, for example, the Chinese National Natural Science Foundation, a new division of funding should be set up to support popsci activities, including book writing.
4. In major research institutes, such as those in the Chinese Academy of Sciences, one-year visiting positions for prospective writers could be established, enabling them to observe the research in action, learn about recent major research findings, and discuss with the experts or perhaps even collaborate with them to write popsci books.
5. Most importantly, to guarantee that popsci books will be sold in large quantities in the immediate future, all science teachers in high schools and universities should incorporate the use of popsci books in their classes. It is done by offering the students extra credit if they buy a popsci book, read it and write a brief report. This is a sure way to excite the students in science and to enlarge their knowledge base. (See item 1 in Section 5.2.)
6. Since natural science forms the basis of all social sciences [Lam, 2008a; Wilson, 1998], and since science and literature are equally important in shaping modern lives, the time has come to include several popsci books—such as James Watson's *The Double Helix* [Watson, 2001]—into the list of required readings in the general education of every student in every university.

In points 1-4, the prospective popsci writer should be allowed to come from any place (especially magazines and newspapers) as long as the candidate is qualified. Naturally, points 5 and 6 are equally applicable to other countries. China is a country with a strong central government and these recommendations do not need that much new funding; they can be implemented quickly. What is needed is the willpower to do so. Luckily for China there is a tremendous amount of willpower, as impressively demonstrated in her organization of *Olympic 2008*.

## 5.5 Education Reform: A Personal Journey

Education reforms in universities could involve any of these three components:

1. Contents of the course
2. The teaching method of the instructor
3. The learning method of the student

No matter how it is done, an unavoidable constraint that will crucially affect the success of the reform is usually not mentioned, or ignored completely by the reformers; that is, *the reform should not increase the teaching load of the instructor*. Also, the quality of the student taking a course—like the quality of a sample in a physical experiment or the raw material in a factory—is of primary importance; this factor is never emphasized enough. Obviously, with a defective sample, no good experimental result can be expected, no matter how skillful the experimentalist is. This last factor points to the need to start any education reform from grade one on, or even better, from the kindergartens. And I am not kidding.

With the constraints understood and resources limited, I tried to do my best as a teacher. There is not much we can do about item 3 above. It is very hard for the student to change her/his learning habit after being wrongfully taught for 12 years before they show up in college, and this is not their fault. I therefore concentrated my effort in the first two items.

On item 2, the instructor's teaching method, I have tried something radically different. It is called "MultiTeaching MultiLearning" (MTML) [Lam, 1999]. We note that in a physics class, the instructor usually does not have enough time to cover everything. The attention span of a student is supposed to be about 15 minutes. Students in a class have different learning styles. Some students are more advanced than others. Active learning and group learning are good for students. Around 1999, to overcome these problems in the teaching of two sections of a freshmen course in mechanics, I have tried a zero-budget and low-tech approach. In this course, we covered about one chapter per week, using *Physics* by Resnick, Halliday and Krane as the textbook. In each course, there were three classes per week, each 50 minute long. In the last session of every week, the class was broken up completely. Different "booths" like those in a country fair were set up in several rooms, manned by student volunteers from the class. The rest of the class was free to roam about, like in a real country fair, or *like what professional physicists do in a large conference with multiple sessions*. In this way, we were able to simultaneously offer homework problem solving, challenging tough problems for advanced students, computer exercises, Web site visits, peer instruction, and one-to-one tutoring to the students. The students seemed to enjoy themselves and benefited from it. However, this approach was soon discontinued. It did require a little bit of extra preparation from the instructor; but more importantly, it did not seem to raise significantly the grades of the students. The "inferior raw material" factor might be at work here.

The next thing I tried, with better luck this time, is to integrate popsci books into my physics classes, as described in item 1 in Section 5.2. This practice was quite successful; the students liked it very much.[16]

This popsci book program is not trying to alter the course content *per se*. My first attempt in this direction, item 1 in education reform above, actually happened earlier. Soon after I started teaching at SJSU in 1987, I created two new graduate courses, Nonlinear Physics and Nonlinear Systems.[17] But these two courses were for physics majors. In Spring 1997, I established a general-education course called *The Real World*, opened to upper-division (that is, third and fourth years in college) students of *any* major. It results from my many years of research ranging from nonlinear physics to complex systems [Lam, 1998]. The description of this course is given in the flyer in Fig. 5.4. There were only nine students in class, majoring in physics, music, philosophy and so on, plus two physics professors sitting in. It was fun. The course stopped after one semester due to nonacademic reasons, falling victim to the sociology of science education.

---

[16] American students are crazy about extra credits in a course, even though the time they would spend to do the extra-credit work could or should be used in learning the course itself. It is a psychological thing, probably frequently used by teachers from grade one on.
[17] These two courses resulted in two textbooks, one for undergraduates [Lam, 1998] and the other for graduate students [Lam, 1997].

Five years later in Fall 2002, the course was resurrected with the same name but modified to suit incoming freshmen students. It is this general-education freshmen course that will be described in detail in the next Section.

---

**A brand new course for students of any major!**

It is time to go beyond textbooks
and learn something about

# The Real World

Phys 196 (3 units), Spring 1997
MW 4:00-5:15 pm

The course contains unified descriptions of the real world, with themes from fractals, chaos and complex systems, and applications in many social and natural systems. In addition to homeworks, the student has one of three options: (i) take a written final exam, (ii) do a report on a popular science book, or (iii) do a project on any topic selected from the daily newspaper. Topics include:

- DNA and information
- Predictions in the financial market
- Traffic problems
- Can one model Darwin?
- "The Bible" and "Gone With The Wind," What is in common?
- What does a computer scientist know about AIDS?
- Why we are here?

**Prerequisite: An open mind.** (No advanced math beyond algebra; computer knowledge not needed, but plenty of chance to use your computer skills if the student so desires.)

**Instructor: L. Lam (Sci. 303, 924-5261, lullam@email.sjsu.edu)**

---

Fig. 5.4. The upper-division course, Phys 196: The Real World, offered in Spring 1997.

## 5.6 The Real World

In 2001 we have a new provost in campus. This very energetic and ambitious man, Marshall Goodman, wanted to make SJSU distinctive among the 20 plus campuses of the California State University system. Introducing international programs with a global outlook was his way of doing that. But perhaps more important, with lightning speed as administrative things went, he was able to push through the university senate and actually had 100 brand new freshmen general-education courses set up and running in about half-a-year's time. Each of these courses is limited to no more than 15 incoming *freshmen* students. The program starting in Fall 2002 was called the Metropolitan University Scholar's Experience (MUSE). Here is the official description of the MUSE program:

> University-level study is different from what you experienced in high school. The Metropolitan University Scholar's Experience (MUSE) is designed to help make your transition into college a success by helping you to develop the skills and attitude needed for the intellectual engagement and challenge of in-depth university-level study. Discovery, research, critical thinking, written work, attention to the rich cultural diversity of the campus, and active discussion will be key parts of this MUSE course. Enrollment in MUSE courses is limited to a small number of students because these courses are intended to be highly interactive and allow you to easily interact with your professor and fellow students. MUSE courses explore topics and issues from an *interdisciplinary* focus to show how interesting and important ideas can be viewed from different perspectives.

### 5.6.1 *Course Description*

"MUSE/Phys 10B (Section 3): *The Real World*," created and taught by me (Fig. 5.5),[18] was one of the 100 incoming-freshmen MUSE courses.

---

[18] I was so enthusiastic about this course that I delayed my sabbatical leave by one semester, from Fall 2002 to Spring 2003, in order to teach it in Fall 2002.

## 1. Course description

To *understand how the real world works from the scientific point of view*.[19] The course will consist of two parallel parts. (1) The instructor will introduce some general paradigms governing *complex systems*— fractals, chaos and active walks—with examples taken from the natural and social sciences, and the humanities. (2) Students will be asked to pick *any* topic from the newspapers or their daily life, and investigate what had been done scientifically on that topic, with the help from the Web, library, and experts around the world. Outside speakers and field trips are part of this course.

Metropolitan University Scholar's Experience
San José State University

PROFESSOR LUI LAM
MUSE Faculty –Physics Department

Fig. 5.5. The plastic card certifying Lui Lam as a MUSE faculty.

## 2. Student learning objective and goals specific to this course

After successfully completing this course, the student will:

- Realize that there are general paradigms—fractals, chaos and active walks—governing the functioning of complex systems in the real world, *physical and social systems* alike.
- What nonlinearity is.

---

[19] SciMat by *design* restricts itself to the scientific study of *humans*; it is thus part of this course which is about *everything* in the universe, as indicated by this statement (and the contents of the course). In turn, histophysics by *definition* is part of SciMat.

- How "dimension" is defined mathematically.
- The meaning of self-similarity and fractals.
- Recognize and able to evaluate data to show that any physical structure or pattern in the real world is a fractal or not.
- What a chaotic system is.
- Able to distinguish a chaotic behavior from a random behavior given the time series of a system.
- To realize that many complex systems in the real world can be described by Active Walks, and be familiar with a few examples.
- Recognize that there are multiple interpretations or points of view on some ongoing, forefront research topics, and that these interpretations can co-exist until the issue is settled when more accurate data and a good theory become available.
- Know the difference between science and pseudoscience, and the real meaning of the *scientific method*.
- How scientific research is actually done.
- Able to find out the latest scientific knowledge about any topic of interest in the future.
- Have improved your skills in communicating both orally and in writing.
- Have increased your familiarity with information resources at SJSU and elsewhere.

3. Course material

The following book is required:

Lui Lam, *Nonlinear Physics for Beginners: Fractals, Chaos, Solitons, Pattern Formation, Cellular Automata, and Complex Systems* (World Scientific, 1998), paperback (list price: $28). Reading assignments from this book will be announced in class. Additional material will be provided by the instructor. Other information could be found from the Web, magazines, research journals and books from the library.

4. Grading

The final grade of 100% for each student is split among several items:

| | |
|---|---|
| Homework | 20% |
| Tests (3 total, including final; 10 points each) | 30% |
| Term project and presentation | 20% |
| MUSE activities | 15% |
| Field trip | 5% |
| Participation | 10% |
| Total | 100% |

A *term project* is required. It is a group effort with three to four students in a group. The topic will be chosen by the group, with the help and consent of the instructor. Progress of project will be presented by group members orally in class throughout the semester. A written progress report is to be handed in about the middle of the semester, and a written final report is due at end of semester.

5.  Teaching philosophy

The class is run like a *research group*, with flexibility in content and timing according to the progress and need of the students, and with the injections of other foreseeable and unforeseeable academic activities. The instructor will teach some basic knowledge about complex systems, while each term-project group will be treated like a research group. Each student will be trained to be a scholar, working individually and as a member of a team.

6.  Topics covered by the instructor

*Part I*

1.  *The World is Nonlinear*
    1.1 Nonlinearity
    1.2 Exponential growth
    1.3 Gaussian distribution (the bell curve)
    1.4 Power laws
    1.5 Complex systems are nonequilibrium systems
2.  *Fractals*
    2.1 Classification of patterns
    2.2 Self-similarity
    2.3 Definition of "dimension"
    2.4 What is a fractal?
    2.5 Fractal growth patterns

3. *Chaos*
   3.1 Sensitive dependence on initial conditions
   3.2 The logistic map
   3.3 A dripping faucet
   3.4 Chaotic *vs.* randomness
4. *Active Walks*
   4.1 What is an active walk?
   4.2 Examples of active walks
5. *Conclusion*
   5.1 Simplicity can lead to complexity
   5.2 Order can arise from chaos
   5.3 The world can be understood scientifically

## Part II

These special topics will be inserted between the chapters in Part I, as time allowed:

- How scientific is the scientific method?
- Science *vs.* pseudoscience
- How research topics are born
- Diversity: The first woman president of the American Physical Society
- Does the world have any meaning?

### 5.6.2 The Outcome

There were 12 students in the class. In the beginning, every student was asked to buy and read a newspaper, pick out the topics that interested her or him, which could be about international conflicts, movies or television programs, sports, or anything. After class discussion, three topics— Creativity, Predictions, and What Is Life?—were chosen. Three groups with four students each were formed; each group focused on one of the three topics. Each group tried to find out the current status and the frontier in the scientific study of the chosen topic—through books, the Web and interviewing of experts. Each group gave regular progress report in class and, at the end of semester, handed in a written report after orally presenting it. Simultaneously, the instructor gave lectures on nonlinear and complex systems (see item 6 in Section 5.6.1). .

At the end, we were all exhausted. The students seemed to have a good time. Did they really get the message that the real world can be understood and is governed by some unifying principles? Only time can tell. But it was a nice try.

My feeling is that this course is better offered to non-freshmen who are more mature and motivated. In fact, this course—with the content and approach intact but the depth of coverage modified—could be taught at any level, for undergraduates or graduate students.

## 5.7 Conclusion

Looking back, ever since I published my first paper on nonlinear physics, on propagating solitons in liquid crystals in *Physical Review Letters* in the year 1982 [Lam *et al.*, 1982] while I was working at the Institute of Physics, Chinese Academy of Sciences, I have been doing research on systems of increasing complexity—from solitons to pattern formation to chaos and to complex systems. After the invention of active walks in 1992 [Lam, 2005b; 2006c] and after 1998, the year *Nonlinear Physics for Beginners* [Lam, 1998] was published, I tried to apply AW to human-related systems, ending with the creation of histophysics in 2002 [Lam, 2002]. From that point on, it was easy for me to enlarge the vision and come up with the idea of Science Matters [2008a], focusing myself on studying humanities from the perspective of complex systems.

The review of my past activities in SciComm and PopSci as well as teaching presented in this chapter makes it clear, at least to me, that my research direction is strongly coupled to and influenced by these activities; *vice versa*. I hope this example will encourage others to try the same. Many of the experiences I went through could be easily borrowed by others, or hopefully would inspire them to innovate, in the interest of SciComm, SciMat and education reform.

At this point, I hope you have found out and understand my two reasons for breaking the rule in writing the title of this chapter. If not, please go back to read item 5 in Section 5.2.

# Appendix 5.1: Popular-Science Books Selected in Classes

Sample lists of popsci books selected in my classes are presented in Tables 5.1 and 5.2 here. (See also [Von Baeyer & Bowers, 2004].)

Table 5.1. Popular science books both chosen and bought by students themselves in a freshmen calculus-based physics class in Spring 2000.

| Title | Author | Year |
|---|---|---|
| The Art of Happiness | Dalai Lama/Cutler | 1998 |
| Beyond Einstein | Kaku/Thompson | 1995 |
| The Big Bang Never Happened | Lerner | 1992 |
| Black Holes, Worm Holes, & Time Machines | Al-Khalili | 1999 |
| A Brief History of Time | Hawking | 1998 |
| Calendar | Duncan | 1998 |
| Clones & Clones | Nussbaum/Sunstein | 1998 |
| Comets | Levy | 1998 |
| Computer | Campbell-Kelly/Aspray | 1996 |
| Darwin On Trial | Johnson | 1993 |
| The Diamond Makers | Hazen | 1999 |
| Faster Than Light | Herbert | 1988 |
| Fuzzy Logic | McNeill/Freiberger | 1994 |
| Fuzzy Thinking | Kosko | 1993 |
| Genesis & the Big Bang | Schroeder | 1990 |
| The Hidden Heart of the Cosmos | Swimme | 1996 |
| Immortality | Bova | 1998 |
| The Little Book of the Big Bang | Hogan | 1998 |
| The Meaning of It All | Feynman | 1998 |
| The Mind of God | Davies | 1992 |
| Night Comes to the Cretaceous | Powell | 1998 |
| 101 Things You Don't Know About Science and No One Else Does Either | Trefil | 1996 |
| The Physics of Star Trek | Krauss | 1995 |
| The Real Science Behind the X-files | Simon | 1999 |
| Relativity Simply Explained | Gardner | 1997 |
| Science, Technology & Society | Bridgstock et al | 1998 |
| Seven Ideas that Shook the Universe | Spielberg/Anderson | 1987 |
| Sex & the Origins of Death | Clark | 1996 |
| Skeptics & True Believers | Raymon | 1998 |
| Skies of Fury | Barnes-Svarney | 1999 |
| Steven Hawking's Universe | Filkin/Hawking | 1997 |
| There Are No Electrons | Amdahl | 1991 |
| To Engineer is Human | Petroski | 1992 |
| The Universe and the Teacup | Cole | 1998 |
| Why the Earth Quakes | Levy/Salvadon | 1995 |
| Why Sex is Fun? | Diamond | 1997 |

Table 5.2. Popular science books selected by the instructor for the students to pick, in the upper-division class of Thermodynamics and Statistical Physics in Spring 2000.

| Author | Title | Year | Remark |
|---|---|---|---|
| H.C. von Baeyer | Warmth Disperses and Time Passes: The History of Heat | 1998 | Story of heat and the scientists involved; Maxwell's Demon; time's arrow. |
| T. Schachtman | Absolute Zero and the Conquest of Cold | 1999 | Story of how scientists lower the temperature; not that exciting, author not a scientist. |
| M. Riordan & L. Hoddeson | Crystal Fire: The Invention of the Transistor and the Birth of the Information Age | 1997 | Very exciting story; shows how good science was done in Bell Labs.; a must read especially if you live in the Silicon Valley. |
| G. Johnson | Fire in the Mind: Science, Faith, and the Search for Order | 1995 | Science and religion near Santa Fe, including studies in information and complexity. |
| A. Guth | The Inflationary Universe: The Quest for a New Theory of Cosmic Origins | 1997 | Written by the inventor of inflationary universe; unique; exciting physics and story. |
| T.A. Bass | The Eudaemonic Pie | 1985 | The story of UC Santa Cruz students, applying what they learn about Newtonian mechanics and chaos to beat the roulette in Las Vegas. |
| W. Poundstone | The Recursive Universe: Cosmic Complexity and the Limits of Scientific Knowledge | 1985 | All about cellular automata, with computer program for Game of Life. |
| J.D. Barrow | The Artful Universe: The Cosmic Source of Human Creativity | 1995 | Power laws, fractals, music. |
| M. Schroeder | Fractals, Chaos, Power Laws: Minutes from an Infinite Paradise | 1991 | Fits our course; highly recommended |

# References

Glover, W. [2000] *Cave Life in France: Eat, Drink, Sleep...* (Writer's Showcase, Lincoln, NE).

Gregory, J. & Miller, S. [2000] *Science in Public: Communication, Culture, and Credibility* (Perseus, Cambridge, MA).

Han, X.-P., Hu, C.-D., Liu, Z.-M. & Wang, B.-H. [2008] "Parameter-tuning networks: Experiments and active-walk model," *Euro. Phys. Lett.* **83**, 28003.

Lam, L. (Lin, L.), Shu C.-Q., Shen, J.-L., Lam, P. M. & Huang, Y. [1982] "Soliton propagation in liquid crystals," *Phys. Rev. Lett.* **49**, 1335-1338; **52**, 2190(E) (1984).

Lam, L. [1997] *Introduction to Nonlinear Physics* (Springer, New York).

Lam, L. [1998] *Nonlinear Physics for Beginners: Fractals, Chaos, Solitons, Pattern Formation, Cellular Automata and Complex Systems* (World Scientific, Singapore).

Lam, L. [1999] "MultiTeaching MultiLearning: A zero-budget low-tech reform in teaching freshmen physics," *Bull. Am. Phys. Soc.* **44**(1), 642.

Lam, L. [2000a] "Integrating popular science books into college science teaching," *Bull. Am. Phys. Soc.* **45**(1), 117. (Also reported in *APS News*, March 2000.)

Lam, L. [2000b] "How Nature self-organizes: Active walks in complex systems," *Skeptic* **8**(3), 71-77.

Lam, L. [2001] "Raising the scientific literacy of the population: A simple tactic and a global strategy," in *Public Understanding of Science*, ed. Editorial Committee (Science and Technology University of China Press, Hefei).

Lam, L. [2002] "Histophysics: A new discipline," *Mod. Phys. Lett. B* **16**, 1163-1176.

Lam, L. [2003] "Modeling history and predicting the future: The active walk approach," in *Humanity 3000, Seminar No. 3 Proceedings* (Foundation For the Future, Bellevue, WA) pp.109-117.

Lam, L. [2004a] "A science-and-art interstellar message: The self-similar Sierpinski gasket," *Leonardo* **37**(1), 37-38.

Lam, L. [2004b] *This Pale Blue Dot: Science, History, God* (Tamkang University Press, Tamshui, Taiwan).

Lam, L. [2005a] "Integrating popular science books into college science teaching," *The Pantaneto Forum*, Issue 19.

Lam, L. [2005b] "Active walks: The first twelve years (Part I)," *Int. J. Bifurcation and Chaos* **15**, 2317-2348.

Lam, L. [2005c] "The origin of the International Liquid Crystal Society and active walks," *Physics* (Beijing) **34**, 528-533.

Lam, L. [2005d] "The story of histophysics: History in the making," presented at *XXII International Congress of History of Science*, Beijing, China, July 24-30.

Lam, L. [2005e] "From history of physics to popular-science book writing: The story of Lillian Hoddeson," *ScienceTimes*, Aug. 4, B1.

Lam, L. [2005f] "The dialogue between science and religion: What to talk about," *ScienceTimes*, Aug. 12, B2.

Lam, L., Li D.-G. & Yang X.-J. [2005] "Why there are no professional popular-science book authors in China," presented at *International Conference on Science Knowledge and Cultural Diversity*, Barcelona, Spain, June 3-6, 2004, *The Pantaneto Forum*, Issue 18.

Lam, L. [2006a] "Science communication: What every scientist can do and a physicist's experience," *Science Popularization*, No. 2, 36-41.

Lam, L. [2006b] "A New concept for science museums," *The Pantaneto Forum*, Issue 21.

Lam, L. [2006c] "Active walks: The first twelve years (Part II)," *Int. J. Bifurcation and Chaos* **16**, 239-268.

Lam, L. [2008a] "Science Matters: A unified perspective," in *Science Matters: Humanities as Complex Systems*, eds. Burguete, M. & Lam, L. (World Scientific, Singapore).

Lam, L. [2008b] "Human history: A science matter," in *Science Matters: Humanities as Complex Systems*, eds. Burguete, M. & Lam, L. (World Scientific, Singapore).

Lam, L. [2008c] "Science Matters: The newest and biggest interdicipline," in *China Interdisciplinary Science*, Vol. 2, ed. Liu, Z.-L. (Science Press, Beijing).

Lam, L., Bellavia, D. C., Han, X.-P., Liu, A., Shu, C.-Q., Wei, Z.-J., Zhu, J.-C. & Zhou, T. [2008] "Bilinear effect in complex systems" (preprint).

Li, D.-G. [2008] "Evolution of the concept of science communication in China," in *Science Matters: Humanities as Complex Systems*, eds. Burguete, M. & Lam, L. (World Scientific, Singapore).

Sagan, C. [1994] *Pale Blue Dot: A Vision of the Human Future in Space* (Random House, New York).

Sanitt, N. [2007] "The First International Conference on SCIENCE MATTERS: A unified perspective, May 28-30, 2007, Ericeira, Portugal," *The Pantaneto Forum*, Issue 28.

Shermer, M. & Grobman, A. [2000] *Denying History: Who Says the Holocaust Never Happened and Why Do They Say It?* (University of California Press, Berkeley).

Shermer, M. [2001] "Starbucks in the Forbidden City," *Sci. Am.*, July.

Snow, C. P. & Collini, S. [1998] *The Two Cultures* (Cambridge University Press, Cambridge, UK).

Von Baeyer, H. C. & Bowers, E. V. [2004] "Resource letter PBGP-1: Physics books for the general public," *Am. J. Phys.* **72**, 135-140.

Watson, J. D. [2001] *The Double Helix: A Personal Account of the Discovery of the Structure of DNA* (Touchstone, New York).

Wilson, E. O. [1998] *Consilience: The Unity of Knowledge* (Knopf, New York).

Yang, X.-J. & Lam, L. [2004] "Research on the history of science in China: Getting hotter," *ScienceTimes*, Aug. 20.

# PART II

# Philosophy and History of Science

# 6

# The Tripod of Science: Communication, Philosophy and Education

*Nigel Sanitt*

Communication in science is not just painted on after the science is finished. It is an essential aspect of both the process and understanding of science itself. This chapter sets out the arguments for enhancing scientists' training to include critical thinking and communication skills, and suggests that all science undergraduates should do at least one philosophy course as part of their curriculum.

## 6.1 Introduction

A hundred years ago the idea that philosophy was not part of a scientist's toolkit would have been greeted with incredulity. In fact the old term for science—*natural philosophy*—portrayed science as a branch of philosophy. So how did the break between science and philosophy come about? And why does the link need to be restored?

The roots of the divide arose out of the prodigious success of science over the last century, which gave rise to a burgeoning overspecialization and fragmentation into different disciplines. University courses became geared up to catering for new areas within science. The "wood" got lost amongst the "trees" and scientific method, principles and critical thinking became jettisoned. Why this trend must be reversed is the subject of this chapter.

The gulf between philosophy and science does not exist, except as one between philosophers and scientists. For those who have been brought up as scientists, many of whom have had little philosophical training, there is a growing need to come to terms with science in a way that can only be satisfied by assuming a more philosophical approach. This is not to say that scientists are ignorant of philosophical issues, but in science courses in the past—especially in England—philosophy was often completely ignored and the result of this was that it simply continued to be ignored by scientists as they progressed in their careers.

The problem also goes beyond science into other subject areas. There is a tendency to view science in a way which is wholly different to other areas of intellectual pursuit. It is often seen as a contrast: science and art, science and literature, science and philosophy. Clearly every discipline has its own agenda, and one has to be careful not to go down the holistic road too far and see everything in terms of a unity which results in meaningless verbiage. On the other hand, the separation between science and other subject areas has certainly been pushed too far. Coupled with this the strong interrelationship between academic disciplines has, as a result, been downgraded and marginalized, with the scientific view ignored or even ridiculed. The popularity of astrology is a sad example of this tendency. But for every action there is a reaction. The need for interdisciplinary research has never been more widely recognized, though this is more often than not accepted in theory rather than in practice. In science experimental data is being amassed all the time and its interpretation is part of the publication process. Debate and analysis continue, usually in tandem, and the interplay between theory and experiment defines the current status of a problem or problems. However, the analysis of the more philosophical aspects of science usually lags well behind, and the intensity of study is in proportion to the perceived difficulty in incorporating a theory within our worldview. My questions are: why should the bulk of this process take place after the event, as it were? And how does one go about incorporating philosophical issues at an earlier stage within science?

Scientists must change the way they think about science, the way they teach science and the way they communicate science. This will only happen if scientists become more thought aware, more skilled in critical thinking and more educated in the philosophical sciences.

## 6.2 Change is Part of Science

The main areas I want to highlight are how science copes with and adapts to change, and how important communication is in science. In particular, I want to stress the multidisciplinary approach, and how within science a lack of philosophical awareness by scientists hampers research at the highest level.

Pressure to change how science is taught is an ongoing problem for universities. Science progresses, new theories come and go, and new courses have to be introduced, often at the expense of material, which is still important. This situation will never change and reflects the fact that science is always in a state of flux.

Change is inherent not only in science but in any subject. Let me explain by taking the example of *Computer Ethics*. In the 1940s Norbert Wiener was one of the founders of modern computing. He wrote *Cybernetics* [1948] and *The Human Use of Human Beings* [1950]. *Cybernetics* (Computer Science or Information Technology—IT) was defined by Wiener as the science of control and communication. Surprisingly, to the first time reader, chapter one of *Cybernetics* is on "Time," with a discussion on Astronomy. The book ends with a chapter on "Society." The importance of society is also reflected in the subtitle to *The Human Use of Human Beings: Cybernetics and Society.*

An ethical side to IT was therefore present right from the foundation of the subject. This was transferred into university courses which involved (and still involve) courses on ethics.

Originally, Computer Ethics was conceived as ethics applied to IT. This was very much in the vein of new twists to old problems. Afterwards, though, there developed a newer broader definition of Computer Ethics, which saw it as a branch of professional ethics. In this new form it was concerned about standards of practice and conduct of IT professionals.

At the present time both definitions apply, and one could say that the field of Computer Ethics has split. This evolution in the way Computer Ethics is thought about is a natural change, which especially affects subjects which are fast growing.

This growth and evolution is not confined just to *soft* sciences, but can also be manifest in the *hard* sciences.

Seyfert [1943] published a list of 12 spiral galaxies, which had very small and very bright nuclei. The nuclei also had peculiar spectral features. This class of objects became known as Seyfert galaxies. In fact, the unusual properties of some of these galaxies had been pointed out by others as early as 1908; however, Seyfert brought this group of objects together and highlighted their importance.

As time progressed astronomers were able to take better quality spectra, obtain x-radiation data, and put together a theory of what was going on in Seyfert and other allied types of galaxies. Nowadays we recognize two types of Seyfert galaxies (rather uninspiringly referred to as type I and II). Out of Seyfert's original list of 12, only 9 or 10 (there is some debate) are now classed as Seyfert Galaxies. Some waggish astronomers have suggested that at some time in the future, none of Seyfert's original list will be classed as Seyfert galaxies!

This has in fact happened to another class of galaxies, which were called *N galaxies*. These were a more extreme class of objects than Seyfert galaxies, but they no longer exist as a class. They have been superseded by a broader class known as *Active Galactic Nuclei*. These latter make up about 2% of all galaxies, and Seyfert galaxies, what used to be N galaxies, and others are all grouped under this category, which reflects the physical processes which they have in common.

The point about these examples is not just that science and the way science is thought about change, but that these changes are philosophically mediated. Observations or experiment involve interpretation and understanding, and there is a complex interplay of attitudes and judgements that motivate and drive evolution. This also occurs in the way scientists think about their subject as well as within the subject.

Thus change is natural both in science as well as how scientists think about science. This does not imply that change is necessary, nor does it explain what changes are needed and how these relate to my first aim, which is to change how science is communicated.

## 6.3 Apathy and Antipathy

Over the last few generations in spite of (and maybe because of) enormous strides in all the sciences, over-specialization and a narrower outlook has crept into science. This has happened while society is demanding better communication with scientists, and just as scientists from different specializations need to communicate better with each other. Unfortunately, the ability of scientists to engage, explain and understand what they are doing is diminishing.

Believe it or not, if that were all that was happening, it would not be so bad. But, there has grown in a small number of scientists an antipathy to all things philosophical (and even some philosophers who would argue it might be better to keep scientists away from philosophy)! Even that can be coped with. What, I believe, is the most damaging problem is a larger group of scientists who are not anti-philosophy, but are simply apathetic towards philosophy. So the situation we are faced with is:

- A broadening intellectual impoverishment.
- Too much over-specialization.
- An apathetic attitude to philosophy, which in some cases is anti-philosophy.
- A communication gap between scientists and society, and between scientists and other scientists.

These are all different aspects of the same problem.

The solution, or at least a part of the solution, is, I believe, to ensure that all science undergraduates do at least one philosophy course as part of their undergraduate curriculum. Ideally, critical thinking/philosophy should be taught much earlier than at the undergraduate level. A report for the Wellcome Trust [Murphy and Beggs, 2005] reported that: "Many children experienced a decline in interest in school science which starts at around the age of 10." This lamentable state of affairs might very well be improved if children were encouraged to think and question more. We still have a long way to go in the UK, to even come close to the experience in France, where philosophy is an integral part of the school curriculum.

Critical thinking leads to better communication in science. Science communication and understanding is not just the delivery of a message, but is an essential aspect of science. It is not just painted on after the science is finished. Of course, neither should science be seen as a problem that needs a solution.

The apathetic attitude to philosophy may appear benign. Consider the view expressed as: "I am a scientist, what has philosophy got to do with my work?" But this attitude is deeply damaging to the research process. Consider the example of a scientist sitting in front of some measurement apparatus twiddling a few knobs. What has philosophy or critical thinking got to do with him or her? The answer is to take a leaf out of Socrates' book and ask a series of questions: Why are you twiddling the knobs? I'm measuring some physical parameter. Why are you measuring this parameter? I'm checking some theory. Explain the theory, why do you think the theory is correct? From the most mundane starting point you do not need many questions in the sequence before you arrive at something close to philosophy, and furthermore, such a sequence will always lead to a philosophical interchange.

Apathy towards philosophy is an untenable position to hold. Philosophical training is a counterweight to ivory tower mentality. Philosophy can provide bridging principles between the sciences and the wider world.

## 6.4  Demarcation

Distinctions between sciences and interfaces between the sciences and other disciplines can be a problem. Consider *Intelligent Design* and Darwin's theory of evolution. To argue the case against Intelligent Design you need to know the arguments. You need to be able to recognize a category error.

Science does not deal with *truth*; if it did, it would be *religion*. It is not a question of Darwin's theory being wrong or right—it is neither: as a scientific theory, it does not aspire to truth—and that is the strength of scientific theories not their weakness.

Alternatively, sometimes crossing the boundary from science to other disciplines can be extremely useful—for example to understanding science. But, you need a map, a thought map, which a thorough grounding in philosophy may provide.

Take *Consciousness* as an example. This is a hot topic at the moment and many disciplines are involved:

- Artificial Intelligence (AI), Computer Science
- Biology, Biochemistry, Medicine and Psychology
- Mathematics and Engineering
- Philosophy and Physics
- Linguistics and Theology

Everyone is plowing their particular fields, but (to be fair) all realize the importance of all the other disciplines. However, you have to be careful to spot when you are trying to compare chalk with cheese.

This multidisciplinary approach in AI has developed in spite of rather than because of pressure from universities. A former director of NASA claimed that he was confident he could lead America to the moon, but he could not change the universities [Ronayne, 1983]. The problem is that departments within universities are rarely multidisciplinary.

The broad interface between science and philosophy also includes cultural aspects of science, representing an area which in the past has been neglected, ignored and even denied. Science is not a technological pursuit within a vacuum; it is a particular kind of intellectually based pursuit which is the result of our interaction within and about the world. There is thus an interplay between science and society, and between science and other intellectual pursuits. At the practical, technological level, science is important in our everyday world; but also at the theoretical level there is a cultural interaction as science affects the way we think and perceive. The cross-pollination of ideas is important, as without it science would become starved of the imagination and brainpower, which it needs. Moreover it is not just philosophy of science to which I am referring but philosophy more generally. Clearly the specific area in philosophy which deals with science is directly relevant and important to science, but this is not to say that it is the only area of importance.

## 6.5  Science Research

I believe that a lack of philosophical awareness and training for scientists affects adversely science research at the highest levels. In order to explain this point, I need to describe the research process methodology. I am not going to embark on a detailed analysis, but it is useful to identify (very loosely) four broad stages of scientific research methodology.

1.  Find a problem

The scientist must strike the right balance between a narrow area that will (hopefully) be fruitful, and a broad area that contains room to maneuver, if initial work proves useless, or leads to a cul-de-sac. The *Goldilocks* scenario of getting things just right can prove quite elusive. Unfortunately, many interesting problems have to be discarded at this early stage. The constraints of time, funding and "delivering results" mean that most research problems are those that the scientist feels can be delivered in months rather than years.

2.  The handle

This entails finding the right entry path (of many) that will result in a successful outcome. When research is successfully completed, it often turns out that there were a number of possible ways into the problem. A researcher does not want to spend an inordinate amount of time on a problem, which could be solved in a different way more simply.

3.  Real science

This corresponds to the actual research work. However, it also involves critical assessment at every stage, problem (often crisis) management and dealing with others.

4.  Writing up

Here communication skills come to the forefront. But even a lone researcher will have to deal with referees, editors and the media.

    I have accentuated the craft aspects of scientific research. The question is: Is it acceptable to learn all this *on the job*, like some glorious amateur? And possibly compromise vital aspects of the scientific enterprise through lack of sufficient training. The philosopher Karl

Popper always claimed that he learnt most of his philosophy when he was an apprentice cabinetmaker, but we can't all go and learn to make desks!

So what can go wrong at the research level, and be put down to bad philosophy? Consider the following example which is taken from the astrophysics of black holes and neutron stars.

## 6.6  Black Holes

First, some background, and I have kept the technical details to a minimum. Neutron stars are thought to be one of the possible end points of evolution of a star. They are approximately 20 km in diameter and have a mass of around 1.4 times that of our sun. They are the remnant that is left behind after a star explodes as a supernova. They are incredibly dense and can occur as isolated objects or in binary star systems as well as at the heart of old supernovae clouds.

Black holes are theoretically postulated entities. They are thought to be the end point of stellar evolution when the remnant left behind after a supernova is too massive to form a neutron star. The upper mass limit for a neutron star is thought to be below 3 times the mass of our sun.

Most stars occur in binary systems. Our own sun does not have a binary star companion, although if Jupiter had been a bit bigger, it would have been a star.

Some binary systems emit large amounts of x-rays. It is thought that in these cases one of the stars in the binary undergoes a supernova explosion, leaving behind a neutron star or possibly a black hole. In the case of binary stars we can measure the masses of the stars involved, and as a result, two classes of x-ray binaries have been identified. These two classes are designated *low mass* (around 2 solar masses) and *high mass* (greater than about 3 solar masses). The two classes are not arbitrary; there does seem to be a grouping of masses around 1.4 solar masses in one group, and a range of masses in the other group from 3 up to 15 solar masses and beyond. Also, in the low mass group, the x-ray luminosity sometimes "bursts." This refers to a sudden increase by a factor of about ten in x-ray output, which then decays back to a quiescent level; the high mass x-ray binaries do not seem to share "bursting" behavior.

The theory behind these observations is that in the low mass x-ray binaries you have a neutron star and in the high mass x-ray binaries you have a black hole candidate. So far so good. The use of the word *candidate*, which reflects that black holes are not proven, is very important. In the case of neutron stars they exhibit pulsations (hence *pulsars*) and their nature is not in question at the present time.

The theory behind the bursts which occur in the low mass x-ray binaries is that in-falling gas accretes, compresses, and heats up via thermonuclear reactions on the surface of the object, which results in sporadic bursts of x-rays.

So what can go wrong from a philosophical point of view, and how can this affect research? The answer to this lies in the fact that whereas neutron stars have surfaces, black holes do not. More precisely, the theoretical entity, which we refer to as a black hole, has an event horizon. Matter falling in towards a black hole is redshifted and appears to be swallowed up at the event horizon but is not compressed, so as to initiate thermonuclear reactions. The event horizon of a black hole is not a physical barrier or surface, but a point (or more precisely a surface) of no return. Thus for black holes, no surface equals no x-ray bursts. This is certainly consistent with the observations, but some authors have gone much further [Narayan and Heyl, 2002]. They claim that the lack of bursts in the high mass x-ray binaries is actually evidence for the presence of an event horizon, and consequently evidence for a black hole.

This is a step too far. Observations can provide arguments consistent with an event horizon rather than a surface for an object, but they cannot prove it. "Absence of evidence is not evidence of absence" [Abramowicz *et al.*, 2002].

Strictly speaking the formal fallacy committed is "denying the antecedent." This is of the following form: "If A implies B, then not A implies not B." In the present case: If "x-ray bursting" implies "object has a surface", then "no x-ray bursting" implies "object does not have a surface." The fallacy is easily seen in the following example of similarly fallacious reasoning: "If I am in London, then I am in England" implies "If I am not in London, therefore I am not in England."

The problem from the research point of view is that a huge amount of futile effort has gone into trying to find evidence for black hole event horizons. From a "sociology of science" perspective, competition between groups has been the driving force, in this instance, for one group to abandon reason.

From the communication point of view it is interesting to note that in a later study Narayan's group claimed "indirect evidence" of event horizons for black hole candidates, which was amended in their press release to "strong evidence" [Narayan, 2005].

The use of the terms "indirect evidence" in a research paper and "strong evidence" in a press release may be considered by some as harmless "poetic licence," and if it helps with funding—so much the better: but it is not an example of science practice at its best.

## 6.7 Communication

The communication of science is inextricably tied in with its practice, education and philosophy. The need to reduce and compartmentalize the world on the one hand, and to go for the holistic approach on the other, is a continual balancing act, which scientists have to perform all the time. Extremes on both sides lead to difficulties. On the one hand, there is the big picture, but on the other, a too simplistic position, which ascribes some single name to an overall complexity, ends up devoid of meaning. There is an infinite amount of information "out there" and you just cannot list it all. From the scientific viewpoint, therefore, trying to achieve this balance always introduces an element of compromise into science communication.

### 6.7.1 *Language*

Communication has many faces; scientists speak in many languages. I do not refer to French or German etc., but to different styles, technical and non-technical, which scientists use. Physicians talk to their patients in a different language style than when they speak to other physicians. Scientists generally have to tailor their language style depending on the audience they are addressing.

Science often introduces concepts which are unfamiliar and even counter to common sense. The scientist has the advantage of being able to utilize mathematics and visual communication aids in the text, and coupled with an acceptable layer of neologisms can try to balance brevity with clarity. A jargon-free non-technical language for science in the end is virtually impossible to attain. In the case of the reader who is approaching a scientific work for the first time some preparation is essential in order to follow the technical language. In fact, learning the language is part and parcel of scientific training.

Science is also often presented in a false historical sequence. In particular, textbooks rarely explain a subject area in strict temporal order. The reason is that the order in which things are explained is important in understanding a subject, even if that order is not the order discoveries were actually made.

### 6.7.2  *Metaphor*

The level at which researchers cross into different fields is normally highly relevant to their work and the "neighbouring" fields are usually perceived as very close. For example, clearly engineering and biology are seen as important subjects relevant to those working in the field of artificial limbs, and chemistry and astrophysics play an important role in connection with work on molecular clouds in space. What is perhaps less clear is the importance of, for example, the use of metaphor in science. Yet for an explanation of the duel character of fundamental particles in quantum theory it may be that this may be a profitable area worth exploring. The problem here is that whereas engineering, biology, chemistry and astrophysics are all seen as "sciences," metaphor is a linguistic term within the realm of literature and thus outside science. The interaction between thought and the world is not that between two sides of a divide. We are part of the world and our thoughts are shaped by the world. It is a question of our human nature and evolutionary history, which by interacting with our surroundings have moulded the way in which we cope with the world.

The use of figurative and metaphorical language "borrows" from daily life. Scientists have to resort to everyday analogies to explain new ideas to the public at large. There are inevitable problems in stretching language in novel ways: "The language of Science is a language under stress" [Hoffmann, 2002].

In the end this can be highly beneficial to the understanding and explanation of science—even poetic! Consider the two following quotations [Wheeler, 1988]:

> There are innumerable clouds of probability running around in the Universe that have yet to trigger some registered event in the macroscopic world.[1]
>
> Can Spacetime be everything—both what *is* and what *happens*?[2]

### 6.7.3 *Getting the Message Across*

The message has to be mediated through journalists as well as scientists. Many journalists have science degrees—some do not. An extreme example was the case of an AIDS researcher at a press conference who knew he was in trouble when a reporter asked him to spell "virus" [Tanne, 1999]. The trouble also is the bandwagon effect when some areas of science become elevated to "Popular Science." Popular science is often bad science [Grace, 2001].

## 6.8 Conclusion

My aim is to show that, in science, philosophy or critical thinking:

1.  Plays a constructive role in analysing and weeding out false or naive ideas, counteracts scientism/dogmatism and promotes understanding of science.
2.  Improves communication—particularly writing, verbal and media skills.
3.  Helps break down barriers between science and society. It emphasizes the cultural aspects of science.

---

[1] See p. 338 in [Wheeler, 1998].
[2] See p. 257 in [Wheeler, 1998].

4.  Humanizes science creating an ethical dimension.
5.  Motivates and guides research.

In society, at the present time, there is a trend to dumb-down science. But following Newton's third law of every action having an equal and opposite reaction, there is also a counter-movement. Society is fascinated and even in awe of scientific developments.

You often hear the cry: "Rocket science is hard." Well, it isn't. Plug the numbers in a computer and it will work out where the rocket is going. What is much harder is dealing with people. You can't plug the numbers into a computer and handle the media, governments, grant-awarding authorities or the public in such a predictable way.

# References

Abramowicz, M. A. *et al.* [2002] "No observational proof of the black-hole event-horizon," arXiv: astro-ph/0207270 v1.

Grace, J. [2001] "Beware the gods of science," *Evening Standard*, London, February 12.

Hoffmann, R. [2002] "Science, language and poetry," *The Pantaneto Forum*, April, Issue 6.

Murphy, C. & Beggs, J. [2005] "Primary science in the UK: A scoping study," *Wellcome Trust*, April, www.wellcome.ac.uk/assets/wtx026636.pdf.

Narayan, R. & Heyl, J. S. [2002] "On the lack of Type I x-ray bursts in black hole x-ray binaries: Evidence for the event horizon?" ArXiv:astro-ph/0203089 v2.

Narayan, R. [2005] arxiv.org/abs/astro-ph/0509758 and universe.nasa.gov/press/event_horizon/event_horizon.html.

Ronayne, J. [1983] "Science policy studies: Retrospect and prospect in science under scrutiny," in *Science under Scrutiny: The Place of History and Philosophy of Science*, ed. Home, R. W. (Springer, New York).

Seyfert, C. K. [1943] "Nuclear emission in spiral nebulae," *Astrophysical Journal* **97**, 28-40.

Tanne, J. H. [1999] "On the one hand, on the other hand: We're all guilty, at least some of the time," *21st C*, Fall, Issue 4.2.

Wheeler, J. A. (with Ford, K.) [1998] *Geons, Black Holes, and Quantum Foam: A Life in Physics* (Norton, New York).

Wiener, N. [1948] *Cybernetics: or Control and Communication in the Animal and the Machine* (Wiley, New York).

Wiener, N. [1950] *The Human Use of Human Beings: Cybernetics and Society* (Houghton Mifflin, Boston, MA).

7

# History and Philosophy of Science:
# Towards a New Epistemology

*Maria Burguete*

This chapter discusses a new concept of doing and thinking about chemistry while this scientific field is engaged with new phenomena. The new concept can give new answers to old problems related to the study of receptors. Whereas modern science gives preference to instrumental and experimental as well as analytical methods, a new era has emerged in the field of chemistry—computational chemistry—from the mid-20$^{th}$ century onward; that is, computer-aided ligand design methodology has been used to study and represent the characteristics of dopamine receptor structures. An explanation for the most significant epistemological approach in the change of notions and in basic phenomena discovered by using the method of computational chemistry (similar to a theoretical/philosophical case study) is given. The philosophical interest of this approach is connected with a different way of looking at chemistry, especially in the case of computational chemistry: It is an *epistemological* approach because it deals with language and classification in chemistry.

## 7.1 Introduction

All the early sciences stemmed from Philosophy. Descartes, Mach, Bohr and Einstein were aware of the philosophical basis of their search for knowledge; before them there was even no distinction at all between science and philosophy. However, from around two hundred years ago the paths of the scientist and the philosopher separated from each other:

1. The philosopher accuses the scientist of daring to philosophize without sometimes knowing the more profound thoughts of Kant, Hume or Wittgenstein.
2. The scientist rebels against the apparent esotericism of much philosophical discourse and when Philosophy of Science is mentioned, she/he is tempted to close the book and exclaim: Here is a person whose ideas would be much clearer if he/she stayed a few months in my laboratory.

The new data that scientists have at their disposal nowadays is sufficiently subversive to call into question the very form taken by traditional philosophical propositions. Thus, there is nothing more natural than to provide philosophers with this new data to reflect upon.

However, the history of scientific research in the 20$^{th}$ century has taught us something that has led to much greater disorder in our inherited ideas, with event after event revealing worlds that no longer have anything to do with the furtherance of our intuition or with our daily habits of thought.

Historical research tends to be a narrative structured in such a way that facts from the past may be understood and presented in a logical sequence. History of science is important because it gives us the knowledge about the discovery of certain facts which allows us a better understanding of science evolution. Inventions of instruments such as the computer together with sources of computer information allowing instantaneous data transfer, have transformed the everyday life of all scientists [Collier, 1989].

Philosophy of science with their philosophical explanations try to overcome the marks of time or place. However, these explanations are not well accepted in the physical and biological professions of today. Part of the reasons for this is certainly the limited knowledge or complete ignorance of Philosophy by most scientists. Limited knowledge can even have worse consequences than complete ignorance. Since philosophers differ widely among themselves, and each historical period has its own philosophical debates, a limited knowledge could mean a strong bias in favor of some philosopher, who happens to be presently in fashion. Jerome Berson [2000] has criticized thus Karl Popper convincingly, showing that if his views had been known by the end of the 19th century, and had been taken literally by chemists of that period like August von Kékulè and his followers, it would have been the death of the spectacular advancement of Organic Chemistry during that period of time. Berson shows that the intellectual constructs used by chemists, although not being testable in the sense required by Popper, guided organic chemists successfully to a coherent and useful construction of that branch of chemistry. Non-chemists are often unaware of how many submicroscopic structures and mechanisms became well established among chemists as explanations for experimentally observed properties and functions of substances, without being verifiable or falsifiable in themselves in a rigorous way. In a more specific manner this has been expressed by Roald Hoffmann [2007], the Nobel Laureate of chemistry 1981, in the following words:

> The reliable knowledge gained of the molecular world came from the hot and cool work of our hands and mind combined. Sensory data, yes, but we did not wait for scanning tunneling microscopes to show us molecules; we gleaned their presence, their stoichiometry, the connectivity of the atoms in them, and eventually their metrics, shape and dynamics, by indirect experiments..... Amazing that one could design the reality of physiologically active pharmaceuticals and billion-ton industrial production on such seemingly flimsy knowledge, isn't it?

Curiosity, seen as expectant behavior in the face of the Universe, has awakened in humankind the passion and the desire to observe and to know more. This in turn has led the human race to succeed as a biological species. Through this process, science has followed its many paths in an attempt to find answers to the ever-increasing number of questions.

## 7.2  Perspectives of Science

We know that subjectivity is always to be found influencing the relationship that grows between the object of study and the researcher who studies the object. Most scientists look at the world through the eyeglass of their own mind and cannot see it with eyes that are not his own.

Science can be looked at from various perspectives, thereby generating a series of questions. Let us analyze the various perspectives of science:

1. The perspective that is interested in the status of theories and their relationship to what is real.
2. The perspective that concerns mainly social mechanisms that promotes both scientific production and dissemination.
3. The perspective that highlights the role of science in the immaterial field which results from strategies of discovery, legitimization and communication.

Here we will focus on the status of theories, namely, the theory of receptor-ligand binding and their implications in view of new achievements in the scientific research of biochemistry and medicinal chemistry. For a better understanding of this subject let us first present a brief survey upon the history of contemporary chemistry.

## 7.3 History of Contemporary Chemistry

Considering the extent that chemical methodology has contributed to other disciplines, it is tempting to take the charge that chemistry is in danger of losing its identity [Corey, 1989], or, turn it around and proclaim[1] instead that chemistry—today more than ever before—is the "central science" (Fig. 7.1).

All those scientific fields outside of the circle in Fig. 7.1 have made their development through chemistry one way or the other. Let us present some practical examples:

1. *Pharmacology*—understanding drug action mechanisms through the mechanisms of the chemical reactions involved.

---

[1] Discussion by R. B. Woodward in [Todd, 1956].

2.  *Medicine*—a better understanding of therapeutic agents at the molecular level made possible the synthesis of new drugs as well as the achievement of new treatments.
3.  *Biology*—the development of molecular biology made possible the de-codification of human genome.
4.  *Materials science*—all materials are made of molecules; with the synthesis of new molecules new materials can also be achieved. Nanotechnology is a very good example.
5.  *Physics*—the real world is made up of molecules; occasionally, the better we understand them the better we understand the world of physics or the physical world.
6.  *Mathematics*—the development of computational chemistry helps to further the advancement of *ab initio* methods[2] which involve a great knowledge of mathematics.

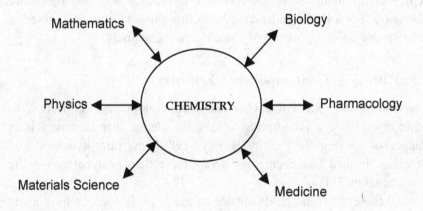

Fig. 7.1. Sketch of Chemistry (including organic synthesis) as the "central science" among natural and technical sciences.

---

[2] The *ab initio* method has been used in the calculation of energy levels and band structures in solid state physics.

Chemistry was given this possibility of acting as "central science" because of the introduction of electronic instrumentation into organic and analytical chemistry research after 1940. It may be seen as a scientific and technological revolution for chemistry in as much as it accomplishes a paradigm shift [Kuhn, 1970]: The replacement of an old paradigm (*analysis by laboratorial techniques* in our case) by a new one (*analysis by computer modeling*).

According to Kuhn's concept, a paradigm is regarded as being a conceptual model used to guide a given scientific activity, implicitly defining both its problems and legitimate methods. The dominant paradigm ensures a pattern of growth, which is eventually overthrown by a "crisis," which is just a psychological concept and not a particular rational cause. And thus the new paradigm appears, and with it a totally new rationality.

Why does a shift in view occur? Genius? Flashes of intuition? Sure. Because different scientists interpret their observations differently? No. Observations are themselves nearly always different. Observations are conducted within a paradigmatic framework, so the interpretative enterprise can only articulate a paradigm, not correct it. Because of factors embedded in the nature of human perception and retinal impression? No doubt, but our knowledge is simply not yet advanced enough on this matter. Changes in definitional conventions? No. Because the existing paradigm fails to fit? Always. Because of a change in the relation between the scientist's manipulations and the paradigm, or between the manipulations and their concrete results? It is hard to make Nature fit a paradigm.

A scientific revolution might seem invisible because paradigm shifts are generally viewed not as revolutions but as additions to scientific knowledge; and because after a new paradigm in a research field is accepted, it is represented in new textbooks.

## 7.4  Paradigm Replacement

Kuhn's research program has gotten a new characteristic: We must study not the mind of individual scientists but that of the scientific community—individual psychology is replaced by social psychology.

Chemists and biochemists especially medicinal chemists, no longer think about synthesis as:

$$A + B \rightarrow C + D$$

where A and B are the raw materials, and C and D are final products. This approach, *Synthetic Analysis*, belongs to the past, wherein, we started from the reagents and following a number of steps from the initial stage till the final stage thus achieving the final product.

Nowadays we think about a certain compound we want to synthesize and look for its precursors from the final product backwards. This new method is the so-called *Retrosynthetic Analysis* which became a reality due to Logic and Heuristic Applied to Synthetic Analysis (LHASA)—a simple yet powerful computer program for the design of organic and bio-organic compounds known as the Logic-Centered Approach [Wipke, 1974]. This approach uses features of the synthetic objective as clues in determining what transformations might be used to obtain the final structure and consequently deduces the structure of the possible precursors.

The first demonstration of this program was held at a Gordon Conference in July 1972 (USA) and then one year later at the NATO Advance Study Institute in the Netherlands, giving rise to Wipke's publication in 1974 mentioned above.

As you can see, from the viewpoint of the chemical community there has been indeed a paradigm replacement with the advent of computer information and computer-based tools useful to the synthetic chemists [Collier, 1989]. More detailed information upon history of contemporary chemistry can be found in [Burguete, 2003].

## 7.5  Philosophy of Chemistry

In the last century philosophy of chemistry has seen some of his old concepts and terminology being changed by several new achievements. Here are two of them.

## 7.5.1  *Transformation Reinforcement Provided by Improved Molecular Representation in Three Dimensions*

Molecules represented in three dimensions (3D) in models gave us another perspective of its reality. This outlook is well explained by Ian Hacking [1983]:

> Experimental work provides the strongest evidence for scientific realism. This is not because we test hypotheses about entities. It is because entities that in principle cannot be "observed" are regularly manipulated to produce a new phenomenon and to investigate other aspects of Nature. They are tools, instruments not for thinking but for doing application.

There is also another aspect that emerges from molecular models: somehow they bridge two worlds with their iconographic representation. On the one hand we can see the the macroscopic world of the compound it represents macroscopic world of the compound it represents but on the other hand we can also imagine the microscopic world that lies beyond its properties and functions. In other words, we could say that iconographic representation turns the invisible visible. In the words of Hoffmann [Grosholz & Hoffmann, 2000]:

> How does the iconic form of the chemical structure expressed as a diagram that displays atom connectivities and suggests the three-dimensionality of the molecule, bridge the two worlds of the chemist? The most obvious answer is that it makes the invisible visible, and does so, within limits, reliably. But there is a deeper answer. It seems at first as if the chemical structure diagram refers only to the level of the microscopic, since after all it depicts a molecule. But in conjunction with symbolic formulae, the diagram takes on an inherent ambiguity that gives it an important bridging function. In its display of unified existence, it stands for a single particular molecule. Yet we understand molecules of the same composition and structure to be equivalent to each other, internally indistinguishable...

Thus the icon (hexagonal benzene ring) also stands for all possible benzene rings or for all the benzene rings (moles or millimoles of them!) in the experiment, depending on the way in which it is associated with the symbolic formula for benzene. The logical positivist in search of univocality might call this obfuscating ambiguity, a degeneracy in what ought to be a precise scientific language that carries with it undesirable ontological baggage. And yet, the iconic language is powerful, efficient and fertile in the hands of the chemist.

### 7.5.2 *Methodologies of Computational Chemistry Provided by Computer-Aided Ligand Design*

Methodologies of computational chemistry (Fig. 7.2) provided by Computer-Aided Ligand Design (CALD) [Burguete, 2006] have shown us that chemistry develops through a special logic which implies a lot of imagination and creative mind, being these two elements not well accepted by logic itself.

Or in other words, we can argue that the distinctive properties of molecules [Golinski, 2003] as well as the functions they accomplished show us the irreducibility of chemistry to physics therefore expressing the distinctiveness of chemistry. In the words of Hoffmann [2007]:

> Chemistry climbs ladders of complexity, create new molecules and emergent phenomena with lots of intuitive thinking on the horizontal level. Making up a story while making molecules.

In CALD we work with models, that can be a structural model, and these models are subjected to computational "experiments" to deduce information about their properties and thereby to find out which hypothetical new structures will have the desired properties.

The seven columns from left to right in Fig. 7.2 represent the following items, one item one column:

1. Molecular graphics and data visualization

Computer graphics appeared in 1971 and is perhaps the most important component of computational chemistry: graphical depiction of molecules has lines representing bonds connecting points representing atoms. Multidimensional properties can be shown on the "surface" of a molecule by color-coding electron-density distribution or electrostatic potential for instance.

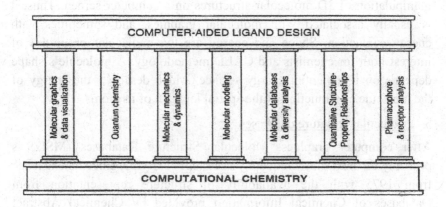

Fig. 7.2. Methodologies of CALD.

## 2. Quantum chemistry

The aim of quantum chemistry is to describe the average spatial distribution of electrons, thereby in the process computing a molecular energy and other molecular properties. The energy can be computed for different nuclear configurations (molecular geometries), thereby allowing what is called geometry optimization. Therefore, one important feature of quantum chemistry in CALD is to optimize structures (to find low-energy conformations).

## 3. Molecular mechanic and dynamics

Molecular Mechanics (MM) and Molecular Dynamics (MD) are the two components of Force Field Methods: the so-called force field is a set of energy terms and associated parameters. The methods of MM are well suited for both small and large biomolecules but provide information about a model that is motionless and has no kinetic energy. In reality, molecules undergo vibrations and at physiological temperature undergo many conformational changes. This brings up the subject of MD which appeared around 1990s.

## 4. Molecular modeling

Molecular modeling is often used to describe the interactive manipulation of 3D molecular structures on a computer screen. Thus, it is usually associated with molecular graphics, and sometimes with energy evaluation. Shape and energy are two molecular properties of interest both for chemists and CALD methodology. A molecule's shape depends on its potential energy surface, which describes the energy of the molecule as a function of the spatial location of its atoms.

## 5. Molecular structure databases

After computer graphics, Molecular Structure Databases (MSD) is probably the next most frequently used tool of CALD, since its existence from 1975 with the availability of structure representation from Databases of Chemical Information provided by Chemical Abstract Service. There are two types of MSD:

(1) The 2D type stores atoms (chemical elements) and connectivity information, for example, which atoms are bonded to one another.
(2) The 3D type stores in addition, the (x, y, z) Cartesian coordinates of each atom in a molecule. In this case it is possible to search for compounds with both connectivity and geometrical specifications.


## 6. Quantitative structure-property relationships

The goal of this class of methods, Quantitative Structure-Property Relationships (QSPR), is to find and quantify relationships between an observed molecular property for a set of compounds and what are called descriptors (descriptors can be almost any quantity that helps to describe a molecule). When the molecular property is biological activity, then the applications are called Quantitative Structure-Activity Relationships (QSAR).

## 7.   Pharmacophore and receptor analysis

In this category, Pharmacophore and Receptor Analysis, are placed those CALD tools that deal specifically with ligand and receptor shape. In this method, called hypothesis generation, software tools are designed to analyze the shapes of a series of compounds that are known to bind the same receptor. A conformational search is done on each molecule in a set chosen to span the bioactivity range from highly potent to barely active and the program tries to find a conformation of each molecule so the more active ones all can achieve the same pharmacophore (pharmacophores are the essential functionalities of a molecule necessary for its pharmacological activity).

The goal is to use the methodologies of computational chemistry to help discover chemical structures with properties that might qualify them to enter the drug discovery pipeline. And this is possible because there is really *intuitive thinking on the horizontal level*. So, there must be application of the computer information to generate ideas for new substances with desired properties.

## 7.6   A Case Study: Functional Selectivity

The fundamental idea behind CALD research is depicted in Fig. 7.3 and the concept behind this diagram is that the nuclear and electronic structure underlies all physical and chemical properties of molecules, including their biological activity. This was the state-of-art underlying CALD methodology until the recent discovery made by Richard Mailman [2004] about a new concept called *Functional Selectivity*.

In Fig. 7.3 we realize that the therapeutic effect is based mainly on the molecular and electronic structure of the molecule as well as its chemical and physical structure. Afterwards we have to consider the nature of the active site responsible for drug-receptor interaction, which in turn induces a biochemical change as well as a biological response from the receptor once it is activated.

Before going further, let us explain a few concepts for those who are not chemists or biochemists:

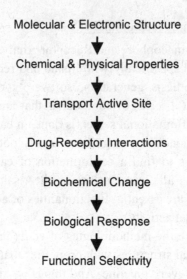

Fig. 7.3. The Computer-Aided Ligand Design diagram.

1. A *ligand* is a small molecule that fits into and binds, usually temporarily, to one of the biomolecule of a living organism, just like a key-lock fitting.
2. A *receptor* can be a biomolecule to which another molecule (the ligand) will bind, thereby evoking a cascade of chemical steps that will produce a physiological effect.
3. A *receptor site* is that specific region of a biomolecule to which a ligand binds.

Receptor sites have the ability to first attract and then bind to a ligand through electrostatic interactions.

Now we are ready to understand that in the diagram of Fig. 7.3 underlying all the other properties a compound can exhibit are its 3D disposition of atomic nuclei and the electronic distribution around the nuclei. These particles of physics determine the chemistry (reactivity and physical properties) the compound can undergo. The properties, in turn, determine how that molecule will interact with other molecules. The interactions determine solubility, lipophilicity, association and stability which in turn will be decisive for efficiency of binding at the receptor site. As a final result of these interactions we will have an *agonist effect* (the receptor is activated) or an *antagonist effect* (there is no receptor activation).

If we have an agonist effect then a cascade of biochemical events occur and will eventually be observable in the patient in terms of therapeutic response to the drug; if we have an antagonist effect there is a non-effective binding therefore no biochemical event will be detectable. For example an agonist of histamine receptor will enhance histamine production when administrated to the patient; an antagonist of histamine receptor will inhibit histamine production.

However, in recent studies made by Richard Mailman in 2004 we are face to face with a new concept—Functional Selectivity, where a single drug can bind to a single receptor and yet cause a mix of effects (agonist, partial agonist, and/or antagonist). This effect is explained by *Function Activity Hypothesis*, which posits that some drugs may have markedly different functional effects through a single receptor. These drugs can cause effects as diverse as pure antagonism versus full agonism. This new hypothesis states that there is a conformational change induced by ligands that may cause differential functional activation at the dopamine receptor site.

Whereas most drugs are ligands only very few ligands are drugs, because even small variations in chemical structure can influence whether the compound will be curative, physiologically inert or toxic. Nevertheless, in a general way, compounds or substructures that possess similar shape, volume, lipophilicity, electronic distribution and others, cause similar effects in a biochemical system [Burger, 1994]. In the words of Richard Mailman [2004]:

> Not only the functional selectivity hypotheses provide an explanation for previous contradictory observations, it also opens new horizons for drug discovery. It provides a route to novel drugs targeting only some of the functions modulated by a single receptor, something with implications at both the basic and clinical level.

*M. Burguete*

To elucidate the molecular mechanisms underlying functional selectivity a combination of approaches including computational, molecular and biophysical/chemical approaches in which both ligand and receptor structure are considered. Computational chemistry provides powerful tools to help chemists to simulate and visualize these molecular properties, as well as to refine raw experimental data on receptor structures. Molecular structure being the universal language of chemists, a user sitting in front of a workstation will draw a molecule to be modeled into the computer using a "point and click" paradigm: the simplest graphical depiction of molecules has lines representing bonds connecting points representing atoms. Multidimensional properties can be shown on the "surface" of a molecule (Fig. 7.4) by color-coding electron-density distribution, electrostatic potential or relative hydrophilic character for instance [Schleyer, 1998].

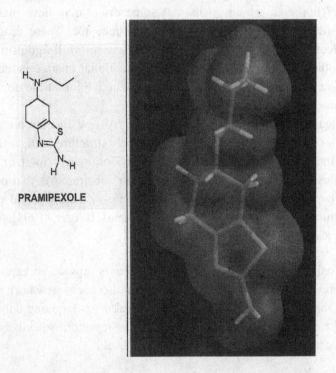

Fig. 7.4. Pramipexole: a selective dopamine agonist

## 7.7 Philosophy of Science and Epistemology

The history and philosophy of science speaks to us of science as both a historical issue and a philosophical issue. Philosophy by definition is the beginning and the end, the first question and the last explanation, the search for the absolute truth about humanity and the universe, both exterior and interior, whilst science is a means, the inquiry and the discovery of partial truths. The task of establishing the essential normative principles of science as a unit and the reflective and higher integration of the various specialized types of scientific knowledge will fall to the Philosophy of Science.

Epistemology arose as a substitute for the Philosophy of Science in 1908 with Émile Meyerson, a chemist and a philosopher. Epistemology deals with a variety of areas:

1. The study in a general, inclusive way of the implications for the universe of the scientific theories developed.
2. It is related to sociological aspects: scientists and their explanations and presuppositions throughout history, theology, determinism and probability.
3. The clarification of new scientific concepts, even if sometimes using old terminology.
4. The analysis of scientific procedures—hypothesis formulation, development of methods, explanation of facts, social applications.

The case study we have presented deals with two specific areas of epistemology:

1. The clarification of new scientific concepts, even if sometimes using old terminology

The old concept of key-lock receptor theory should be replaced by the new concept of functional selectivity because now it is known that a receptor site no longer exists in one conformation only; it can assume different conformations therefore generating the possibility of multiple receptor sites instead of only one receptor site which led to the old pharmacological dogma of key-lock concept.

2.  The analysis of scientific procedures—hypothesis formulation

As a consequence of the new concept of functional selectivity we now have *function activity hypothesis* (that is, some drugs may have markedly different functional effects acting through a single receptor) which provides the possibility of mix effects, agonists and antagonists, instead of just one particular effect separately. The old hypothesis that postulated that a molecule could only act as an agonist or an antagonist is no longer valid in light of this new hypothesis.

The Philosophy of Science in the $20^{th}$ century is a scientific philosophy of the sciences, which adopts rigorous criteria and a methodological process that are analogous to the work of science itself. In Science, you do; in the Philosophy of Science you think how to do, what for and why.

## 7.8  Conclusion

There are two questions that functional selectivity has raised.

1.  Does nuclear and electronic structure underlies all physical and chemical properties of molecules including their biological activity?
2.  How a *single* drug can bind to a *single* receptor and yet cause a mix of effects (agonist, partial agonist, inverse agonist, and/or antagonist)?

The answer for these questions is still under study yet some of the conclusions are:

1.  At the scientific level

We can no longer say that therapeutic effect is based mainly on the molecular and electronic structure of the molecule as well as its chemical and physical structure. A single drug can bind to a single receptor and yet cause a mix of effects such as agonist and antagonist effects.

## 2. At the epistemological level

In this new epistemological approach—*take everything by its interface value*—computer graphical interfaces overcome a culture based on calculus giving rise to a new culture: *a culture of simulation.*

We are no longer looking at chemistry only from the vertical understanding (the classical reductionism method) where we started from raw materials and then achieve the final product. Horizontal understanding is expressed in the concepts and symbolic structures at the same level of complexity as the object to be understood. As Hoffmann [1988] said, *"...horizontal explanations are quasi-circular."* Horizontal explanations are systemic.

Reality can be seen as a process of information exchange among several levels (biological, physical, psychological and social). Modern scientific knowledge is based on the concept of relationship or interaction, a much broader concept than the concept of analysis still used by normal science. Therefore, science became epistemic.

Take everything by its interface value means that computers encourage a new style of thinking—*tinkering* [Turkle, 1999]—what essentially means empiricism. Interactive mode is the *modus operandi* of modern computers, therefore creating the dominant style of the actual thought: systemic thinking.

The study of visualization and representation is another philosophical issue that has come increasingly to the fore with the development of chemistry and the parallel growth of computational power. Chemists are rather unique in frequently needing to visualize structures and entities that they also know not to exist according to the dictates of physics, such as molecular orbitals.

On the way to a new epistemology!

# References

Berson, J. A. [2000] "Kekulé scapes, Popper notwithstanding,"*Angew. Chem.,* Int. Ed. **39**(17), 3045-3047.

Burger, A. [1994] "Medicinal chemistry: The first century," *Med. Chem. Res.* **4**, 3-15.

Burguete, M. C. [2003] *Arquitectura de Novas Moléculas: Uma Abordagem Dinâmica,* Ph. D. Thesis (Fundação Calouste Gulbenkian & Fundação para a Ciência e Tecnologia, Ministério da Ciência e do Ensino Superior, Lisbon).

Burguete, M. C. [2006] "The philosophy of computational chemistry. II," in *Proceedings of the 2nd ICESHS: The Global and the Local: The History of Science and the Cultural Integration of Europe,* ed. Kokowski, M. (Cracow, Poland).

Golinski, J. [2003] *Chemistry,* The Cambridge History of Science, Vol. 4, Eighteenth-Century Science (Cambridge University Press, Cambridge, UK) pp. 377-396.

Collier, H. R. [1989] *Chemical Information: Information in Chemistry, Pharmacology and Patents* (Springer, Berlin).

Corey, E. J. [1989] "Something valuable from almost nothing: A personal view of synthetic chemistry," *Chemist* (Washington, D.C), July/August, 3-5.

Grosholz , E. R. & Hoffman, R. [2000] "How symbolic and iconic languages bridge the two worlds of the chemist: A case study from contemporary bioorganic chemistry," in *Of Minds and Molecules: New Philosophical Perspectives on Chemistry,* eds. Bhushan, N. & Rosenfeld, S. (Oxford University Press, Oxford, UK).

Hacking, I. [1983] *Representing and Intervening* (Cambridge University Press, Cambridge) p. 262.

Hoffmann, R. [1988] "Nearly circular reasoning," *American Scientist* **91**, 9-11.

Hoffmann, R. [2007] "What might philosophy of science look like if chemists built it?" *Synthèse* **156** (3), 321-336.

Kuhn, T. S. [1970] *The Structure of Scientific Revolutions,* 2nd Ed. (University Chicago Press, Chicago) pp. 43-52.

Mailman, R. [2004] "Functional selectivity: Novel mechanism of drug action," *Med. Chem. Res.* **13**(1-2), 115-126.

Schleyer, P. (ed.) [1998] *Encyclopedia of Computational Chemistry,* Vol. 1 (Wiley, New York) p. 799.

Todd, A. R. (ed.) [1956] *Synthesis: Perspectives in Organic Chemistry* (Interscience, New York) pp. 155-184.

Turkle, S. [1999] *Predictions: 30 Great Minds on the Future* (Oxford University Press, Oxford) p. 329.

Wipke, W. T. [1974] "Computer-assisted three-dimensional synthetic analysis," in *Computer Representation and Manipulation of Chemical Information,* ed. Wipke, W. T. (Wiley, New York) pp. 147-174.

# 8

# Philosophy of Science and Chinese Sciences: The Multicultural View of Science and a Unified Ontological Perspective

*Bing Liu*

Many debates and a few new perspectives focusing on Chinese sciences appeared recently in China. Here, after some historical and philosophical ideas are reconsidered, a unified ontological perspective is proposed to explain the existence of and the rationality behind the multicultural view of science.

## 8.1 Recent Debates on "Chinese Sciences" in China

Recently within China, there are many debates on Chinese sciences. For example, whether Chinese medicine or feng-shui is a kind of science, and debates on pseudoscience. A typical example is that a Chinese philosopher of science wrote a paper to advance "Farewell to Traditional Chinese Medicine and Remedies" [Zhang, 2006]. He argued that Chinese medicine does not belong to the medical sciences and is not exactly a rational medicine. He went on to view Chinese medicine from the perspective of "culture progress," "respecting science," "species diversity maintenance" and "humanitarianism." His paper has provoked a big debate about Chinese medicine, which is still going on in China. In this debate, an academician of the Chinese Academy of Sciences even wrote a paper to advocate that the core theory of Chinese medicine—yin-yang and the five elements—is nothing but pseudoscience. Opinions like these are typical in the recent debates on Chinese medicine.

Feng-shui is the topic in another debate. Many people certainly regard feng-shui as a typical pseudoscience or superstition [Tu, 2006]. Similarly, in the last three years in the debate on "reverence to Nature" (referring to ecological environment and the relationship between human and Nature)—not quite encouraged by the government, words like "reverence to Nature is anti-science, even anti-human beings" are advanced.

Tracing back to earlier years, the Chinese academic circle was interested in discussions like "whether there was science in ancient China" and the "Needham Question" [Needham, 1969], which actually was related indirectly to these debates in some sense, because the recent debates refer to how we think about "science," especially how we think about "Chinese science" and how it differs from the present Western science. Hiding behind these debates is a deep ideology background, such as "scientism," "Western science centralism," "monism science view," "the concept of truth in China," and so on.

## 8.2 The Multicultural View of Science

The foregoing debates are actually related to how we consider the problem of demarcation in Western philosophy of science. Could a new boundary of science arise from an analysis outside of the scope of traditional Western philosophy of science?

For example, two questions have confused scholars in the history of science and have not been settled: the question of "whether there was science in ancient China," and the Needham Question. Early in the 20th century, some Chinese historians of science such as Ren Hongjun and Zhu Kezhen had discussed the former question [Ren, 1915]. In fact, the science they referred to is the system existing in modern Europe, including scientific theory, experimental method, scientific organization and criterion, etc. If we define science like this, science of course did not exist in ancient China. However, those who base their argument on national pride and so forth reach the opposite conclusion. Otherwise, how could one justify the study of history of science in ancient China if there is no science in that period of time?

Recently, remarkable emphasis on the concepts of "local knowledge" and "cultural relativism" and the corresponding "multicultural view of science" in "nontraditional" philosophy of science [such as sociology of scientific knowledge (SSK), feminism, postcolonialism and scientific anthropology] offer an inspiring approach—pluralism in science concepts—on foregoing questions. In the history of science, traditional research concentrates only on modern Western science, which is rooted in a monistic view of truth and science. With the development of postmodernism, this kind of monistic view was deconstructed, resulting in the emphasis of local knowledge, and the appearance of a new, wider, multiple view of science.

Within this new conception system, since the universal, single, normative science system no longer exists, and since modern Western science is also a local science system, we need not take it as a standard any more but can simply legitimize the study of history of Chinese sciences, all by itself. The Needham Question can be resolved in the same manner. There is no need to discuss such question like "Why modern science did not appear in China?" if both Western and Chinese sciences are local knowledge.

Let us go back to the debate on the Chinese sciences, taking *Chinese medicine* as an example. We can see that if we define science wider and take a multiple view of science, and assign all humans' systemic cognitions of Nature to the conception of science, Chinese medicine (and other medicines, such as Mongolian medicine, Tibetan medicine, Indian medicine, etc.) can then become what we call "science." Rather than to define the concept of science superficially, it is meaningful to ask whether the main stream approach of modern Western science (and its corresponding "paradigms") is the only way for us to cognize and understand Nature. The fundamental basis of modern Western science is empirical validation[1] while Chinese medicine, as a local knowledge, has gone through a long-time evolutionary selection in its development based on its effectiveness.

---

[1] Of course, empirical validation may not be entirely objective because observations are always laden with theoretical concepts.

Following these thoughts, many problems in the debate about Chinese and Western medicines could be understood anew. For example, in the past, there are at least three viewpoints about Chinese medicine:

1. Those against Chinese medicine view it as a "superstition" or "pseudoscience."
2. Those who adopt Western science as the only acceptable concept emphasize that Chinese medicine is still a true "science."
3. Some others want to put Chinese medicine into the theoretical framework of Western medicine.

However, all these three views have their root on "scientism," resulting from the acceptance of modern Western science as the standard reference. Following these views, Chinese medicine would never be treated respectably and would not be developed properly in the future. Consequently, it is imperative to change our basic idea of what science is.

Thus, the knowledge which was formerly regarded as folk belief, witchery or superstition has its legitimate status in the multiple view of science. For example, acupuncture therapy, herbalism, ancient maternity, woman physician or heeler, etc., deserve the same attention as given to traditional knowledge (such as mathematics, astronomy, physics and so on). Noticeably, although the mainstream studies of science history do include the topics of phlogiston, alchemy, etc., these themes are out of the mainstream of science and are studied traditionally only as a background material relating to science. In short, non-mainstream knowledge has its independent, legitimate status in the multiple view of science.

## 8.3 Lessons from the Study of Art and Science

With the traditional monistic view of science removed, we still need to ask how the concept of multiple sciences (as a kind of formal description) relates to its research object, namely, Nature itself. In our habitual thinking, the pursuance of the ontological basis in philosophy arises from our bred-in-the-bone psychological desire for it. These questions will be discussed below. But first, let us talk about the work from a new research field—the study of Art and Science, which may be revelatory regarding our discussion.

There are differences between art and science in terms of their nature and epistemology. For example, in the simplest description, one considers art as a kind of expression of human beings' inner feeling. Science, on the other hand, regards art as a kind of cognition of exterior nature. With the cultural development of human beings, both art and science became mature and professionalized; separation and division between the two occurred. The division of labor enables art and science to develop in their own ways, resulting in the separation of human pursuit into two different domains: art puts the "laws" away in the pursuit of beauty, while science covers beauty when pursuing the "laws." Consequently, there exist many differences between art and science in understanding Nature: their objects of attention, methods of cognition and expression, techniques of work, criteria for success, and so on. These differences make the communication and interaction between artists and scientists infrequent, even making them isolated from each other.

Furthermore, in addition to the correlation between art and science mentioned above, the two actually share some "sameness" in characteristics. A very important "sameness" is "parallelism" in the way of cognition in art and science, which have been recognized by others. "Parallelism" means that although the methods of cognizing the world by artists and scientists are different, the expression of their "works" are also very different, and they explore the world according to their particular paradigm in their separate fields, they do sometimes obtain similar "conclusions" or similar understanding of the essence of the world. For example, Shlain's famous book *Art and Physics: Parallel Visions in Space, Time and Light* [1991] focuses on how artists and scientists reach the same goal by different routes in the themes of time, space and light. Shlain proposes that both revolutionary art and visionary physics are investigations on the nature of reality. He [Shlain, 1991] says:

> Despite each discipline's similar charge, there is in the artist's vision a peculiar prescience that precedes the physicist's equations. Artists have mysteriously incorporated into their works features of a physical

description of the world that science later discovers.... The artist, with little or no awareness of what is going on in the field of physics, manages to conjure up images and metaphors that are strikingly appropriate when superimposed upon the conceptual framework of physicists' later revisions of our ideas about physical reality. Repeatedly throughout history, the artist introduces symbols and icons that in retrospect prove to have been an avant-garde for the thought patterns of a scientific age not yet born.

Shlain goes on to present systematically some examples of "parallelism," such as innocent art and nonlinear space, barbarism art and non-Euclidean space, fauvism and light, cubism and space, futurism and time, surrealism and relativistic distortion, and so on. In particular, cubism and space is the typical example to express this "parallelism."

In his monograph *Einstein, Picasso: Space, Time and the Beauty* Miller [2001], another American scholar and a science historian, pays attention to the relationship between cubism in painting and the concept of time in science (especially in physics). He presents a contrastive biographical study on the physical scientist Einstein and the artist Picasso, which in fact is a "parallel biographical study." According to Miller, a parallel study will lead to the inevitable question on how art and science developed parallel to each other in the 20$^{th}$ century. From the intellectual endeavor of Einstein and Picasso, we can conclude clearly that art and science were in fact developed in the way of "parallelism" in the last century.

## 8.4 An Ontological Perspective on the Multiple View of Science

Up to this point, we have discussed the similarities and differences between art and science. We also discussed "parallelism" which formally could lead to the appearance of some of these similarities. We now turn our attention to the origin of the "parallelism" itself. Why parallelism exists? What does it mean? In answering these questions, we find that the differences and similarities between various "local" sciences can be understood in terms of the multiple view of science. This outlook is inspired by the parallelism between the cognition in art and the cognition in science.

For exploring the outside world, people could reach the same goal by different approaches. They often obtain essentially the same result by using independent but somewhat similar paradigms. In the multiple view of science, all kinds of "local" science can be coexisting while keeping their own identity. More importantly, in many cases, they may overlap partially and only partially with each other. Accordingly, by considering the mainstream Western science and non-mainstream but culturally related sciences (such as Western medicine, Chinese medicine and other kinds of medicine) altogether, one reaches a picture of multiple sciences. Such a picture brings new thoughts on many interesting problems; for example, the definition of scientific truth, uniqueness of scientific truth, meaning of objectivity of science, etc. When such integration is extended to other human knowledge (art, for example), many more applications are possible.

Here, an illuminative idea should be mentioned. More than 50 years ago Sarton [1962], the founder of history of science and an American historian, suggested that:

Religions exist because men are hungry for goodness, for justice, for mercy; the arts exist because men are hungry for beauty; the sciences exist because men are hungry for truth. The division is not as clear cut as that, but it is sufficient to mark out main oppositions. Think of a triangular pyramid; the people standing on they come nearer as they climb higher. Bigots, little scientists, mediocre artists may feel very distant from one another, but those religion is deeper feel very close to the great artists and the great scientists. The pyramid symbolizes a new kind of trinity culminating in unity.

Sarton's idea is represented in Fig. 8.1.

In the following, we will discuss the unification of science, art and religion. However, at the beginning, we might as well borrow the "model" suggested by Sarton in the sense of ontology and extend it to form our own "model." According to our "model," within science, different theories of science could co-exist with each other and, together, make up the scientific cognition of Nature. Putting it in the pyramid picture suggested by Sarton, all these sciences should be on one side of the pyramid only—the science side. (Of course, the existence of multiple sciences means that science side of the pyramid is not flat at all.) When we include the dimensions of art and religion, regarding these two as ontological objects, the pyramid picture then contains three varieties. Among these three, the cognitive parallelism of art and science is represented by whatever relationship between the art side and the science side of this pyramid.

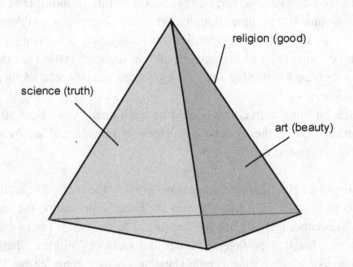

Fig. 8.1. The Sarton pyramid: the three stand-up sides represent religion, science and art, respectively.

At this point, we can imagine extending Sarton's pyramid to a polyhedron (Fig. 8.2) which could have an infinite number of sides corresponding to the infinite dimension for different knowledge of human beings. In other words, in philosophy, we can imagine that the objects, or the ontologically objects of the world, should be an infinite, complex entity which could be represented by such a polyhedron. And each side of that polyhedron corresponds to one knowledge system of human beings. When simplified to a finite one, it becomes a finite polyhedron, with different sides corresponding to different cognitive fields. In this picture, since different cognitions correspond to different sides of this polyhedron (in fact, for one big field each side could be further divided into many sub-sides) they differ from each other, and represent different local knowledge. However, these different sides are connected with each other, directly or indirectly; in this sense, the different cognitions are unified. As far as all local knowledge aim at the same ontological object and all are based on the experience of human beings, all kinds of scientific theory could be "objective." We can find differences among them because they correspond to different sides of the ontological polyhedron. We can find sameness among them because all these different sides belong to one polyhedron and are connected with each other. Then, in the sense of ontology, our knowledge about the world is indeed monistic and is unified as a whole!

Fig. 8.2. The polyhedron picture: each side corresponds to a different view of the same ontological object, situated at the center of the polyhedron.

However, it should be remembered that this is just a modeling metaphor, which, in some sense, is similar to what is described by the Chinese proverb: a blind man feeling an elephant. Even though the whole shape of the elephant could not be grasped by the blind man by touching the elephant locally, yet, the reality of the part touched by the blind man is real and is a part of the whole real elephant.

Our ontological perspective can explain not only why different principles have their similarities and differences, but can also deconstruct the monistic standpoint of objectivity and truth. When we understand the relationship between science(s) and the objective Nature this way we of course will not, as someone else did, put some traditional "sciences" and local knowledge into the category of pseudoscience. Some arguments about Chinese sciences mentioned in the beginning of this chapter could then be deconstructed.

# References

Miller, A. [2001] *Einstein, Picasso: Space, Time and the Beauty* (Basic Books, New York).

Needham, J. [1969] *The Grand Titration: Science and Society in East and West* (George Allen & Unwin, London) p. 45.

Ren Hongjun, [1915] "The reason for why there is no science in China," *Science,* No.1.

Sarton, G. [1962] *Sarton on the History of Science,* selected and edited by Stimson, D. (Harvard University Press, Cambridge, MA) p. 16.

Shlain, L. [1991] *Art and Physics: Parallel Visions in Space, Time and Light* (Harper Perennial, New York ) pp. 1-6

Tu, Jianhua [2006] "Why feng-shui is a pseudoscience?" *Science and Atheism,* No.3, 23-24.

Zhang, Gongyao [2006] "Farewell to traditional Chinese medicine and remedies," *Medicine and Philosophy (Humanistic and Social Medicine Edition),* No. 4, 14-17.

# 9

# Evolution of the Concept of Science Communication in China

*Da-Guang Li*

The history of the conceptualization of science communication (or "science popularization") in China can be classified into four stages. Apparently, the discussion of science popularization evolved from diversified to governmental instructed and to authoritative, and ended up as the goal of raising the public scientific literacy. The last stage is triggered by the worries of the low percentage of scientific literacy in China when compared with that of other countries, as measured in recent years.

## 9.1 Introduction

In the 15th century, Christian missionaries introduced science and technology to China. Defeat of the Qing Dynasty in the Opium War made Chinese intellectuals and the public at large to realize the importance of scientific development and industry.

In the late Qing Dynasty and during the New Culture Movement (1900-1920s), Chinese scientists and intellectuals began pushing the themes of "science and democracy" and "human rights," introducing the ideas of freedom and democracy, and the achievements of natural science from Western countries to the Chinese people.

In 1915, the first science organization was established by a group of scientists who had studied science and technology in the universities of USA. With the publication of the journal *Ke Xue* (*Science*), they wanted to introduce science and technology not merely to the scientists but also to the Chinese public. In 1932, the Guo Min Tang Administration organized the China Association for Scientific Culture (CASC), which disseminated the knowledge of science and technology *via* its journal *Scientific China* and radio broadcasting. CASC had more than 10 sub-organizations and societies in major Chinese cities.

In 1950, right after the founding of modern China, the Chinese government established a professional organization (National Union for Science Popularization) for popularizing science. In 1977, the enterprise of science popularization recovered after the Cultural Revolution (1966-1976). And on December 4, 1994 *People's Daily* published an important editorial to spur up the communication of science to the public, with the strategy of delivering different scientific knowledge and using different method to reach targeted audiences of farmers, youths and cadres (that is, government officers). By mid-2002, the world's first "law of science popularization" was issued in China. In 2006, the China Association for Science and Technology, China State Department of Science and Technology and another nine state departments jointly issued "The Grand Scheme of the Advancement of Scientific Literacy of Chinese People."

Looking back, the history of science popularization in China could be generally divided into four stages:

1.  The late Qing Dynasty and the New Culture Movement period (late 19$^{th}$ century to early 20$^{th}$ century).
2.  The Republic of China period: non-governmental organizations for science popularization (1914-1949).
3.  The period of science popularization with multi-purposes (1949-1994).
4.  The period of long-term objective for the advancement of scientific literacy (1994-2006).

## 9.2  Late Qing Dynasty and the New Culture Movement Period (Late 19$^{th}$ Century to Early 20$^{th}$ Century)

In the 16th century, Western missionaries commenced to spread science to China. Defeat of the Qing Dynasty in the Opium War (1840-1842) made Chinese intellectuals to recognize the importance of science and industry in building China into a powerful country that could defend herself from invaders and get off poverty. In late Qing Dynasty and with the New Culture Movement, Chinese intellectuals introduced the concept of science, natural rights and democracy developed in Western countries to China. They began communicating these ideas and the corresponding knowledge to the Chinese people by way of publishing journals and delivering public lectures.[1] One of the major events was the publication of Yan Fu's *Tian Yan Lun* (*Theory on Natural Evolution*, 1898), translated from Thomas Henry Huxley's *Evolution and Ethics* (1893).

The theory on natural competition and evolution shocked the Chinese; the idea of "natural selection, survival of the fitness" were widely talked about and were even taken up as a topic for composition in primary and middle schools.

## 9.3  Science Popularization by Science Organizations (1914-1949)

In this second stage, the first two important science organizations were established with the tasks of advancing scientific research and science communication. The organizations were copies of their Western counterparts such as The Royal Society established in England, except for their purpose of introducing scientific and technical knowledge and rational thinking to China for eliminating poverty and embedding Chinese culture with scientific thoughts.

In 1915, Chinese oversea students in Cornell University of USA established the Science Society of China (SSC), to rebuild their motherland with the help of science and take up science communication as their principal objective. In China, this organization established a number of laboratories and institutes, and published the journal *Ke Xue* which lasted until 1960.[2]

In 1932, a group of Chinese scientists and government officers established CASC. This association published a journal *Scientific China* (1932-1937). In addition, in major cities throughout the country *via*

---

[1] *New Youth*, Vol. 7, No. 1, Dec. 1, 1919.
[2] *Ke Xue*, Issue 1, Jan. 1915.

broadcasting, journal articles and various activities, this group communicated to the public science issues such as scientific culture, science education, astronomy, military affairs, agriculture, scientific life style, the science spirit, scientific method, etc.[3] The CASC was the first organization with the singular purpose of popularizing science to scientists and intellectuals working for the universities, institutions and the government.

These organizations started the dissemination of science and technology in China. They discussed popular science existing in Western countries; recognized the necessity of introducing the concept of science popularization; defined the terminologies and the concept of science popularization that fitted the local context of the Chinese situation at that time.[4] They also made valuable contributions to the theory of science communication and introduced the practices that could be applied to that "contemporary" period between the end of the Qing empire system and the chaotic society before the War of Resistance Against Japan.[5]

### 9.3.1 *The Early Period*

On June 10 of the 3rd year of Min Guo (1914), a handful of Chinese students at Cornell University—including Ren Hongjun, Zhao Yuanren and Hu Mingfu—decided to set up a science society, SSC, to transplant Western science and technology to China, not only the knowledge but also the "spirit of science, scientific method and the practical knowledge of scientific discovery and the applied sciences." They established the journal *Ke Xue* in January of 1915, trying to "communicate the essence and the impact of science" and "the latest information of the world."

In 1915, after Thomas Alva Edison (1847-1931) was told of *Ke Xue*, he said, "The great nation has waked up." In the 1940s, the famous science historian and professor at Cambridge, Joseph Needham (1900-1995), valued the importance of this journal as a major science magazine,

---

[3] "The declaration of the movement of scientific culture of China," *Scientific China*, Issue 1, Jan. 1, 1933, p. 1.
[4] "What is scientific culture?" *Scientific China*, Vol. 5, No. 8.
[5] "Program for the second stage of the Association for Scientific Culture in China," *Scientific China*, Vol. 5, No. 5.

considering it in parallel to *Science* from USA and *Nature* from UK. The SSC also established several other journals such as *Science Illustrated* (1933), *Journal of Collection of Theses* (1933-1936), *General Discussion on Science and the Issues*, *Selective Journals of Science*, *Selective Books of Science* and *Thesis of Biological Study*.

Ren Hongjun, one of the founders of SSC and the editor of *Ke Xue*, has been considered the most important figure not only in creating the path for introducing science and scientific culture in China but also in contributing heavily to the conception of the science spirit, which he valued higher than science itself as Chinese were concerned. He pointed out five characteristics of the science spirit: truth, accuracy, observation, careful definition and skepticism. That would be more complete if "unafraid of difficulties" and "not greedy for temporary benefits" were included.

Ren Hongjun considered the value of science as the most important in science communication. He thought that the Chinese society was only interested in the benefits of science—that only technology should be brought to the people and thus one should place technology first. He criticized this consideration as "research being only for benefits, not for exploration itself" [Ren, 1916]. He thought that was the reason China had been so weak for such a long period in history, and that was why the people were "either illiterate or flattering." And that was why Chinese science had been so difficult to develop.

Zhu Kezhen, a scientist in xerography, was another one who made much contribution to the concept of science popularization in China at that time. At the fourth conference of SSC on August 15, 1919, his talk was on "The Enterprise of the Science Society of China." He said, "The 20th century is a period that material civilization will be created. Any nation that wants to be competitive must be scientific." At another meeting in 1929, he ventured that "modern science is like a flower grown in a good environment. The environment is the scientific mind of the citizens and it is a very difficult aim to reach." The scientific mind lied in "the attitudes towards science and the scientific method." "Chinese society is far less scientific. It is not easy to change that."

### 9.3.2 *The Late Period*

The Movement of Chinese Scientific Culture was launched in 1932 by CASC, whose members were government officers and scientists. That was the first special organization with the one purpose of disseminating scientific knowledge. The organization administered a semi-monthly journal *Scientific China*, which stirred up a wave of science popularization. The CASC played the role in even deeper exploration of the relationship of science with multi-facets such as social development, economy, national defense, traditional life and even ethics.

The CASC was established in Nanjing, Jiangsu Province. There were only about 50 members when it was first set up. In 1933, *Scientific China* was published to communicate scientific culture movements in China. In the first issue, it announced its aims as being "for research and communicating scientific development in the world; explain and develop traditional Chinese culture with the principles of science to establish a scientific culture in China."

The CASC conducted activities to communicate scientific and technological knowledge as well as scientific culture *via* radio broadcasting (China Central Wireless Radio Broadcasting Station) and sub-societies in major cities all over the country. The scientific and technological knowledge they talked about was closely related to the national economic development, national defense, daily life, traditional culture, etc. The leadership of the organization announced its goal of reaching at least five million people. Unfortunately, up to 1937, only 3,000 people were exposed to the information they disseminated, when Nanjing was occupied by the Japanese troops. The other reason they failed to achieve its goal was due to the fact that a large number of Chinese citizens at that time were illiterate and uneducated.

However, research and discussion by the Chinese intellectuals contributed importantly to the concepts of scientific literacy and science popularization even though they failed to achieve its ambitious goal of spreading them. An article on the first issue of *Scientific China* explained the aim of the organization: encouraging natural science and applied science researchers to communicate scientific knowledge to the citizens and make science the common wisdom of the population. They even highly expected to "create a strong power to get our nation survive the risk and danger it faces and recover the Chinese culture that is declining." "Therefore, we declare the beginning of a scientific culture in our nation."

The objective of scientific culture was "to make science popular and society scientific." Zou Shuwen,[6] a professor of agriculture at Dong Nan University, explained the concepts in an article: "(1) classifying and explaining the traditional culture inherited from our ancestors; (2) educating all the people to understand science; (3) creating our future with the science spirit."

In "The Program for the Second Phase," the scientific culture was further explained:

1.  The attitude to Chinese traditional culture refers to the classification, summing up, explanation and extension of the culture in the past, for the use of modern society.
2.  Effectively communicate scientific knowledge and methods to the people for solving problems of national defense, production and daily life. Turn scientific knowledge into very simple and easily understandable information to educate the public. Scientific knowledge and methods should not only be understood and shared, but to become an effective tool for people participating in defending the country.
3.  Chinese people should be educated and informed of the continuing development of science and technology. All the youths should be educated with scientific knowledge and methods, changing the new generation into a different one equipped with systematic thinking, organizing abilities, correct attitudes and capacity for quick action. Only in this way, can the nation catch up with the development of science in the world and be prosperous in the future.

The CASC placed the communication of military science at the important position because China was just in the War of Resistance Against Japan. The President of the Kuo Min Tang (Chinese Nationalist

---

[6] *Scientific China*, Issue 1, Jan. 1, 1933, pp. 4-11.

Party) that was temporary in power at that time, wrote a few words to encourage the communication of national defense knowledge in the special issue on national defense of *Scientific China*.

The CASC was the first organization with the government's "instructions on the advancement of public understanding of science" included in its policy. The establishment of the association was closely connected with the science policy of the administration. When Kuo Min Tang came into power in 1927, it laid down the foundation for the development of science and technology; for example, the establishment of the Central Research Academy, which built the Zi Jin Shan Planetary Observatory in 1928, reinforced the university education, and organized scientists and technicians to construct railways and bridges across the Qian Tang River.

In addition, the Movement of Scientific Culture of China was an action plan to advance the public understanding of science, with Chinese scientists trying to turn the concept into practice. This movement was the first science wave pushed by Chinese scientists who conscientiously followed the government's instructions on scientific culture. But among the scientists, the discussion on and the conceptualization of scientific culture, scientific literacy, science communication and the way to popularize science within the Chinese social context were absolutely independent and free. In 1937 Japanese troops invaded Nanjing; the movement stopped when the city was occupied. However, what the scientists did for science popularization; their idea that science communication should keep pace with scientific research, with both being developed simultaneously; and their insistence that science values and scientific culture should be embedded in traditional Chinese culture left deeply an enlightening influence on the development of concepts and practices of science communication to come.

## 9.4 Science Popularization under the New Government of Modern China (1949-1994)

In 1949, with victory led by the Chinese Communist Party achieved the People's Republic of China was established. All science activities and science popularization were put under the control of government administrations. During this period, two events were important in the history of science communication within China. In 1949, the Chinese National Federation of Natural Sciences and the China Association for Science and Technology Popularization were established under the suggestion of the Science Society of China, Natural Science Society of China, Chinese Association of Scientists, and Association of Natural Scientists of Northeastern China.

The period from 1949 to 1958 was characterized by the development of diversified conceptualization of science communication according to the instructive ideas from the government and the Party, set in line with the state's interests and policies. In 1958, China Association for Science and Technology (CAST) was established. At its first conference, the President of CAST, Nie Rongzhen, declared the absolute control of science and technology by the Party. Individual membership was replaced by organization membership, to be financially supported by the government [Nie, 1958].

In 1950, Article No. 20 of the Constitution of the People's Republic of China stated that "the nation develops natural sciences and social sciences, popularize scientific and technological knowledge, and encourage science research and technological invention."

Article No. 43 of the Program of the Consultative Conference of the People's Republic of China[7] declared: "To develop natural sciences to serve the industry, agriculture and national defense; to encourage scientific discovery, technological invention and science popularization."

Accordingly, the Bureau of Science Popularization was established under the State Department of Culture with four divisions and one office. The director of the bureau was Yuan Hanqing, a chemist, transferred from the Science Education Museum of Gan Su Province. It seemed that China would start a new movement of science popularization.

Unfortunately, a series of political movements started and run from 1956 to 1976 (Movement of Suppressing Counter-Revolutionaries; Movement of Struggle Against Corruption, Waste and Bureaucracy; Great Leap Forward, and Proletarian Cultural Revolution). Almost everything stopped during the Cultural Revolution for 10 years. The

---

[7] Special Issue on National Defense, *Scientific China*, Vol. 1, No. 8, p. 1.

formal educational system including primary, middle and higher educations stopped. All schools and universities were closed; all students were forced to leave their institutions and go to the countryside to receive re-education by peasants. All organizations including CAST were forced to be dismissed. Scientists and intellectuals had to go to the countryside to "learn from the workers, peasants and soldiers." The whole nation was put in a dark and was frozen for more than 10 years.

## 9.5  Boom of Science Popularization (1994-2006)

On Dec. 5, 1994, an editorial from the *People's Daily*, issued in the names of the Chinese Communist Party and the State Council, was titled "Some Suggestions on Reinforcing Science Popularization." [8] The editorial stated: "Science popularization and education, leading the people to a scientific life and work habit, are the key to change our economic construction—towards the development of a new economic strategy, to be developed by science and technology—and to increase the quality of the labor force. Our contemporary task and the foundation for the development of science and social stability are: advancing the socialist civilization both materially and spiritually." A worry could be sensed in the article: "Some of the superstition and illiterate activities increased; anti-science and pseudoscience appeared more and more often. Those activities have been going against modern civilization, and ruining the thoughts of the citizens and deceiving the masses—harmful to the younger generations and being an obstacle to the progress of society. Therefore, effective measures must be taken to reinforce science popularization. This task is very urgent." The term "scientific literacy" was first mentioned in this authoritative document. It was defined to mean "scientific knowledge, scientific method and scientific thinking," which could be used to "combat superstition, illiteracy and poverty." And "scientific literacy will be much helpful as the foundation of modern socialism is concerned."

---

[8] "Suggestions on the development of science popularization in China," Editorial of *People's Daily*, Dec. 5, 1994.

The Law of the People's Republic of China on Popularization of Science and Technology, the world's first and the only one even now, was issued on June 29, 2002. Scientific literacy was again stressed in the first article of the law.

In January of 2006, the Scheme for the Advancement of Scientific Literacy of the Chinese Public was announced after an almost two-year long research (with the participation of several institutions and universities) was finished. This long-term project set the goal of increasing the population's scientific literacy to match that of Western countries at the level of the 1980s, as measured by the indicators so popularly accepted in the world. The project defined scientific literacy as "the understanding of basic scientific and technological knowledge, basic command of scientific method, awareness of science thoughts and loving the science spirit; and the ability to use them to deal with practical problems and participate in decision making of public affairs."

## 9.6 Conclusion

From the late Qing Dynasty and the New Culture Movement till the founding of New China in 1949, the conceptualization of science communication diversified and organizations (such as the Science Society of China and the China Association for Scientific Culture) were mimicries of The Royal Society based in England. Chinese scientists and intellectuals tried to introduce the essence of science, scientific method, innovative spirit of scientists, the mechanics and systems for the development of science research, and the science spirit to their country fellows. They criticized the outdated social system and called for all the intellectuals to open the door to the outside world, to learn from the advanced experiences of the science systems existing in Western countries. They wrote articles, talked through radio broadcasting and set up volunteer organizations all over the country, to communicate scientific knowledge and make the Chinese citizens to be aware that only science and humanities could save the country. What happened in this stage presented us with the model for active and diversified discussion and debates in the conceptualization of science communication.

After 1949, the Chinese government and the Chinese Communist Party have set up organizations to mainstream science communication and manage discussions. Discussions and activities stagnated and even stopped during a series of political movements. The ten-year (1966-1976) long Cultural Revolution brought a disaster to the development of science communication as well as other fields.

From 1994 to 2006, the Chinese government issued a number of documents and even a law to encourage science communication, and tried to do the communication effectively by sophisticated methods, aiming for targeted audiences such as government officers, farmers, youths and industrial workers.[9] The ambitious long-term plan, Scheme for the Advancement of Scientific Literacy of the Chinese Public, viewed scientific literacy as benefiting the development of national competitiveness and, for the citizens, achieving a better life and gaining the ability to participate in decision making of science policies. Within this stage, the science popularization systems and the management of the corresponding organizations had become more and more government oriented. The conceptualization of science communication, and discussions on the effectiveness of different approaches as well as finding the way to refine the systems for the diversified Chinese cultures had been instructed from above by the government with authoritativeness. The goals of science communication had been kept in line with not only economic development but also social stability. The China Association for Science and Technology had confidentially tried to put the project for developing public scientific literacy into practice, in conjunction with other 14 state departments.

## References

Nie Rongzhen [1958] "The path to the development of science and technology in China," *Red Flag*, No. 9, 4-15.
Ren Hongjun [1916] "On the spirit of science," *Ke Xue* **2**(1), Jan.

---

[9] Law of the People's Republic of China on Popularization of Science and Technology, issued on June 29, 2002.

# 10

# History of Science in Globalizing Time

*Dun Liu*

Although globalization is quite a fashionable word today, from the standpoint of macrohistory, the migration of human civilization may be traced back to the departure from Africa of the early *Homo sapiens*; while the modern tendency towards globalization was initiated in as early as the 15th century when the great geographic discoveries were made. Similarly, although it was not until the first half of the 20th century that the history of science became a really mature discipline, historical records for a certain branch of natural knowledge and mathematics can be paralleled with relevant intellectual products of early human beings. Adopting dialectic narration, this chapter deals with some hot topics in today's historiography, including globalization, history of science, the Needham Question, the C. P. Snow Thesis and cultural diversity.

## 10.1 Globalization Today and Globalization in History

In recent years "globalization" has become a fashionable term, referring to the restructuring of the world economic order especially since the 1990s as a result of the onslaught of capital expanding globally. People in favor of globalization believe that it will lead to the blurring of national boundaries and profoundly influence the way people in every corner of the world live.

What is certain is that the ongoing globalization process will inevitably affect the spiritual life of humankind, impacting on different cultures in new ways and profoundly reshaping them. Globalization is an objectively existing phenomenon confronting the human race, as economic links and the interdependence of countries and regions become ever more significant. No country or region, not even one, will be able to exist independently of the evolving globalized economy.

Nevertheless, there might be some alternative views of globalization. If we may say that the migration of human beings is as old as human history, then the history of globalization may be traced back to the departure from Africa of the early *Homo sapiens*. Likewise, one may argue that the modern tendency towards globalization was initiated in as early as the 15<sup>th</sup> century when the great geographic discoveries were made.

In the past decade, scholars have proposed that globalization mainly occurred in the economic arena and its motive originated from the West. In contrast, someone could say that the globalization process not only encompasses exchanges in economy and trade but also involves the spread of ideas and cultures; and the flow is bidirectional or multidirectional, not only from the West to the East but also the other way around. Contributions to this historical movement of globalization include the trade in the Mediterranean Sea, the three primary religions, the Silk Road, the geographical discoveries, the expansion of colonial rule, the industrial revolution and the two World Wars.

Recent studies of the history of science and technology have established that industrialization and scientific revolutions have promoted the diffusion of scientific knowledge and technology worldwide. In addition, they have narrowed geographical and conceptual distances between people, and expanded the means of human exchanges and cooperation. All these historical processes have resulted in the advent of economic globalization. To put it simply, science and technology have played a pivotal role in the evolution of globalization. Therefore, the history of science might be an excellent approach to understanding the nature of globalization in a profound way. To better appreciate the authentic characteristics of globalization requires a better understanding of science and technology in the past and of their interactions with social advancement.

In the meantime, while some people view globalization with optimism, it does not mean that many cultural differences and traditions that make up the international community are to be ignored. One of the important factors that must be kept in mind is the phenomenon of cultural diversity. This can be compared with the diversity of different species on earth and their perpetual vitality; the world is greatly enriched by such diversity. Likewise, cultural diversity encourages vigour and vitality, as it fosters respect for history and appreciation of the diverse varieties of human cultures and civilizations.

Differing traditions should and do coexist, complementing each other with their merits and contributing to the entirety of human civilization in their respective ways. We are confident that just as humankind will continuously advance, history will not come to an end. The extent to which cultural traditions and local histories are valued will be an indicator of how well a country or an ethnic group is able to develop successfully in the wake of globalization. On the other hand, neglecting or abandoning traditions will render a country lifeless and put it at a great disadvantage in the international arena, as the history of the past one hundred years amply testifies. See Beijing Declaration [IUHSP/DHST, 2005].

## 10.2  History of Science as a Discipline and History of Science as Knowledge

The view that history of science is the knowledge of a specific subject can be traced back to the early days of human civilization. For example, the records on the evolution of Greek mathematics and physics can be attributed to the written descriptions by Eudemus (ca. 370-316 BC) of the development of pre-Euclidean geometry, and that by Plato (ca. 427-347 BC) and Aristotle (384-322 BC), among others, of time, space, movement and force. These were appropriate histories of mathematics, astronomy and physics for their time.

Some exceptional scientists have written on the history of particular disciplines. For example, since 1680, both Newton (1642-1727) and Leibniz (1646-1716) tried to compile the history of calculus, but their objectives lay chiefly in competing for priority of invention, rendering their historical researches unreliable. Since the Enlightenment, a number of professional scientists have shown interest in writing and compiling the histories of different scientific disciplines. Some cases in point include *Histoire des mathematiques* (1758) by the French mathematician

Jean Étienne Montucla (1725-1799), *The History and Present State of Electricity* (1767) by the British chemist Joseph Priestley (1733—1804), and many works on the history of physics by the Austrian physicist Ernst Mach (1838-1916).

Today, although many scientists still are concerned with historical themes, history of science written by scientists is no longer a mainstream trend in the development of the discipline of history of science. It is generally acknowledged that this discipline originated at the end of the 19th century, with its institutionalization completed in the West by the beginning of the 20th century.

When science in some European countries was professionalized in the 19th century, scientists argued that science ought to occupy an important place in the educational system and that more support and funding should be made available to science by society. To this end, they used an effective weapon—the history of science! Driven by this motivation, the authors would normally draw a clear line between scientific knowledge and other forms of cognitive knowledge in their argument that science is a progressive enterprise, while humanistic studies were a side issue, thus asserting that social development and cultural prosperity are a result of scientific advancement. One such representative work is *History of the Inductive Sciences* (1837) written by William Whewell (1794–1866), the former president of the Royal Society in England.

When George Sarton (1884-1956) started promoting the institutionalization of history of science, this discipline in his eyes was by comparison more important and significant than that of religion and of art. In his opinion, only science history can be the proper means of reflecting the advancement of human civilization, and that only scientists, engineers, technicians and medical doctors in each and every field could be counted as *bona fide* makers of history [Liu, 1999a].

Here, the term "institutionalization" refers to structured scientific communities of a certain scale, such as professional publications, a series of academic conferences, specific incentive programs, authoritative academic organizations, acknowledged leaders or schools, corresponding education systems, establishments geared towards the cultivation of successors of the field, etc. This discipline reached maturity in the early part of the 20th century, with landmark events such as the establishment in Belgium by Sarton of the journal *ISIS*, the debut of courses on science history at the University of Harvard in 1920, the funding in 1928 of the International Academy of History of Science, the convention of the first International Congress on the History of Science in 1929, the establishment of a Ph.D. Program devoted to science history at Harvard University, and so on.

This maturing of the discipline saw changes in terms of structure, objectives, content, methodologies, as well as the flavor of research. Works such as studies by John D. Bernal (1901-1971) on the sociology of science, research on the correlation between science and Protestant ethics by Robert Merton (1910-2003), theories concerning Scientific Revolution proposed by Thomas Kuhn (1922-1996), and works by people such as Joseph Needham (1900-1995) on science in non-Western civilizations, have hugely advanced the theoretical basis and pragmatic development in the field. The objective in studying science history does not lie in the simple recording of the evolution of the subject matter. Instead, the focal point is in the interaction between the development of science, technology and society. It is the purport of contemporary historians of science to study science in history within its social, ideological and cultural contexts.

Historians of science of the 21st century will take it as their mission to understand comprehensively the occurrence of science and its growth, and to promote the integration of scientific culture and humanistic culture. Correspondingly, the reason for writing on the history of science is not merely to inform readers of the evolution of science history, but more importantly to equip the public with a balanced cultural literacy and a knowledge system befitting our time [Liu, 1999b].

## 10.3  History of Science in China

China is a country with its own particular tradition of historiography; an unbroken written record of its official history has been kept, which can be traced back to 841 BC. Initiated by Si-ma Qian (ca. 145 BC), father of Chinese historiography and author of *Shi Ji* (*Historical Records*, compiled in ca. 100 BC), some aspects of this tradition, such as *Lü* (musicology and metrology), *Li* (calendar making), *Tianguan* (astronomy and astrology), *Hequ* (hydrography and water conservancy) and *Pingzhun* (transportation and trade) can be compared with the modern concept of "science and technology," and have been included ever since in all official Chinese dynastic histories.

Following this tradition, some scholars paid special attention to historical topics in various subjects in the first half of the 20th century. Nevertheless, compared with the situation in European countries and the United States in the corresponding period, studies on the history of science in China were just in their preliminary stage.

If we evaluate the early works by Chinese scholars from today's perspective, there were some inevitable weaknesses and limitations:

1. Most of these works only concern China, with no consideration for its neighbors in Asia and the rest of the world. Thus, one could say that these early Chinese works on history of science were Sinocentric.
2. Virtually all of these works are devoted mainly to one specific subject, and are neither general nor comprehensive—so it can be said that these are primarily internal rather than external histories of science.
3. Most of the authors were scientists—which raises the question of histories written by scientists as opposed to those written by historians.
4. Last but not least, these works reflect a lack of systematic research and in general are not well organized. In a word, they stagnate at the stage of material collection and rational reconstruction.

History of science as a discipline in China made a late debut; it was not until 1956 that the institutionalization of the discipline commenced, and it was not until 1957 that a national-level institute for the history of science was founded.

However, the institutionalization of science history in China differs from that of its Western counterparts, in that the former was organized from the top down without first acquiring sufficient societal recognition. It is also highly concentrated; that is, administrative decrees were employed from the very outset to concentrate qualified professionals of the field into an organization affiliated to the Chinese Academy of Sciences. With regard to its academic objectives, before institutionalization was achieved, the Research Office for the History of Natural Science, inaugurated on January 1, 1957, stated that her mission was to "summarize the legacies of Chinese sciences, sum up good experiences that the laboring masses have accumulated in the past, and enrich the treasure trove of world science." The allocation of personnel also ensured that the history of ancient Chinese sciences was to be the Office's primary research field [Liu, 1999a].

Consequently, early works on the history of science in China demonstrated two tendencies: first, to describe and promote ancient Chinese sciences; second, to exploit the past for the good of today.

Despite that, by today's standard, these kinds of studies in history of science could not sufficiently embody the essence of the discipline, some first rate works were produced in due time. For example, within the first tendency, high-quality pioneering work in the history of ancient Chinese mathematics was done by Li Yan (1892-1963) and Qian Baocong (1892-1974) [Li & Qian, 1998]. As for the second tendency, Chinese historians of science also made great achievements. One representative work was due to Zhu Kezhen (1890-1974), a meteorologist and former Vice President of the Chinese Academy of Sciences, who made ground-breaking contributions in the study of climate change based on historical records and archives in ancient China [Cho, 1926, 1973]. Another excellent work was Xi Zezong's (b. 1927-) *New Catalogue of Ancient Novae* (1955), which has received wide-spread attention in the contemporary astrophysical community, especially among radio-astronomers [Hsi, 1955].

However, after more than 50 years of endeavor, the institutionalization of science history in China is now on a healthy track, with the following manifestations:

1. A number of doctorate and master training programs have been set up in universities and research organizations, and a batch of historians of science have been trained in both the natural sciences and humanities.

2. In the curriculum policy on natural sciences promulgated by the Academic Degrees Committee of the State Council of China, history of science has been defined as a Grade One discipline, which awards such doctorate degrees as natural sciences, engineering, medicine, agriculture, etc.

3. Since 1999, organizations of history of science at the university department level have mushroomed in a number of institutions of higher learning, such as the Department of History and Philosophy of Science at Shanghai University of Communications, and the Department of History of Science & Technology and Archaeology at the University of Science and Technology of China, in Hefei. In other universities, especially those that are research oriented, research centers relating to history of science have also flourished. A widespread community of the history of science in China has now developed.

4. State leaders and relevant decision makers have given a certain degree of attention to the discipline.

5. Research on history of science has won some societal recognition through the following means: publication of professional journals such as *Studies in the History of Natural Sciences* by the Chinese Society for History of Science and Technology, the employment of a variety of mass media, publication of a quantity of monographs and reference books, awards for books and papers on the subject matter, among others.

6. There has been sufficient interest from international counterparts, and, with the expansion of international exchanges and reinforcement from young scholars returning from overseas, internationalization of the history of sciences and the development of this discipline is taking off.[1]

The continuing growth of the Chinese economy in the past ten years or so has presented a golden opportunity for the advancement of Chinese science and technology. Therefore, research on history of science has

---

[1] D. Liu, "Why history of science should be concerned and supported?" [Liu, 2006].

very valuable bearings on this most populous developing country. For instance, where will the science of this country be in the next 50 years? We may benefit and learn from the experience of the advanced nations in their own development in science and technology; we could also benefit from lessons learned by neighboring developing countries and regions. What is more important is that we bring to mind the path Chinese science and technology embarked upon in the past 50 or even several hundred years, the advancement of scientific knowledge worldwide throughout history, the evolution of scientific structures, the establishment of science policies, the shift of scientific centers, etc. Then we may be able to come up with some kind of model that can be used as a reference for the future development of science and technology in China in a holistic manner [Liu, 2000a].

## 10.4 The Needham Question

In the Preface to Volume 1 of his monumental work *Science and Civilisation in China* (*SCC*), Joseph Needham [1954] proposed a series of questions which he attempted to discuss:

> Why should the science of China have remained, broadly speaking, on a level continuously empirical, and restricted to theories of primitive or medieval type? How, if this was so, did the Chinese succeed in forestalling in many important matters the scientific and technical discoveries of the *dramatis personae* of the celebrated "Greek miracle," in keeping pace with the Arabs (who had all the treasures of ancient western world at their disposal), and in maintaining, between the 3rd and the 13th centuries, a level of scientific knowledge unapproached in the west? How could it have been that the weakness of China in theory and geometrical systematization did not prevent the emergence of technological discoveries and inventions often far in advance (as we shall have little difficulty in showing) of contemporary Europe, especially up to the 15th century? What were the inhibiting factors in Chinese civilisation which prevented a rise of modern science in Asia analogous to that which took place in Europe from the 16th century onwards, and which proved one of the basic factors in the moulding of

modern world order? What, on the other hand, were the factors in Chinese society which were more favourable to the application of science in early times than Hellenistic or European medieval society? Lastly, how was it that Chinese backwardness in scientific theory co-existed with the growth of an organic philosophy of Nature, interpreted in many differing forms by different schools, but closely resembling that which modern science has been forced to adopt after three centuries of mechanical materialism?

Many of Needham's writings concerning the scientific achievements of ancient China are devoted to these and related issues. Numerous answers, including several from Needham himself, have been proposed over the past half century. Generally speaking, the *Needham Question* may be expressed in two ways: Why did modern science develop only in the West after the 16th century? Or, why was the Chinese ancient and medieval civilization more efficient than the West in applying natural knowledge to practical technology and invention?

It is commonly acknowledged that the Needham Question first appeared in *SCC* and his other related works, such as *Great Titration* (1969). In fact, before that, some people had raised similar questions. As early as the 17th century, the Jesuits who came to China had already noted the "backwardness" problem in Chinese science, and consequently in the 18th century, some European thinkers and scientists had tried to find appropriate explanations. Moreover, during the first half of the 20th century, a number of Chinese scholars devoted attention to the "backwardness" issue. All these arguments emerged before the impact of *SCC* on academic circles.

When the Needham Question is brought to the table, such themes as "China" and "science" cross peoples' mind. In particular, it seems that current Chinese scholars have been inclined to treat the "backwardness" issue as an equivalent of the Needham Question. As a matter of fact, the Needham Question has become a widely-discussed topic among historians of science throughout the world, and its full implications go well beyond the more specific matter of "science" and "China," being closely related to hot topics widespread in contemporary academic communities such as "scientific revolution," "modernity" and "cultural diversity."

In fact, the ultimate goal of Needham's *SCC* was to promote mutual understanding among different cultures. The fundamental contribution of Needham's *SCC* is generally deemed to lie in pioneering the integration of non-Western traditions and achievements into world history of science. In a word, science is the common heritage of all humankind. As Needham [1967] pointed out:

The standpoint here adopted assumes that in the investigation of natural phenomena all men are potentially equal, that the oecumenism of modern science embodies a universal language that they can all comprehensibly speak, that the ancient and medieval sciences (though bearing an obvious ethnic stamp) were concerned with the same natural world and could therefore be subsumed into the same oecumenical natural philosophy, and that this has grown, and will continue to grow among men, *pari passu* with the vast growth of organisation and integration in human society, until the coming of the world co-operative commonwealth which will include all peoples as the waters cover the sea.

The problem is: Did the Chinese or any other non-Western nation experience something that we could call "modernization" when they were still unfamiliar with what was normally considered modern science? In developing countries, people should not only explore the reasons for any "backwardness"; more importantly, they may also need to find a way of maintaining the coexistence of modern science and traditional science, and promoting their prosperity together.

Can people really find a way of keeping harmony between humankind and Nature, science and society, development and the environment, global economic integration and cultural diversity? This is a crucial issue for humankind in the new epoch. In this sense, the Needham Question will continue to evoke divergent responses from different parts of the world [Liu, 2000b].

## 10.5  The Snow Thesis and Conclusion

In 1959, British scientist and novelist C.P. Snow (1905-1980) delivered a famous speech at Cambridge University, which was subsequently published as the book *The Two Cultures and the Scientific Revolution* (1960). In his speech, Snow suggested that there were two completely different cultures in the world. His main arguments can be described as follows: Because scientists and scholars of humanities differ in their educational background, disciplinary training, research subjects, as well as in methodology and tools, among many other factors, their basic cultural concepts and value judgments are often in oppose to each other; people of the two camps despise each other to the extent that they do not even try to understand the other's viewpoint. This phenomenon has come to be known as the *Snow Thesis*.

When human society undergoes great transformations in its political and economic structures like that today, it may often result in an era in which cultural conformation and new academic paradigms take shape. The culture of the 21$^{st}$ century should be a new one that integrates fully scientific and humanistic cultures. In other words, the success or failure of the new cultural conformation and academic paradigms is to a very large extent dependent on our understanding of the Snow Thesis.[2] Moreover, in the upcoming cultural development, history of science will play an important role. Not only is this due to the academic characteristic of this discipline in bridging the sciences and humanities, but also because of a more important determining factor; that is, the unique position this branch occupies in human knowledge and the attributes of its research subjects.

Giambattista Vico (1668-1744), an Italian thinker who was the first to give history the status enjoyed by science, in his *Scienza Nuova* (1725) declared the following opinion: Mathematics is artificial and therefore knowable; however, it does not reflect reality; Nature is created by God and therefore beyond complete human comprehension, but reflects reality. Was there a new science that could be fully appreciated by

---

[2]  See [Lam, 2008] for a new discussion of the Snow Thesis and the Needham Question.

humans and, at the same time, reflect reality? He found the answer in Greek mythology and history where humans could communicate with gods. The French philosopher, Auguste Comte (1798-1857), inherited this idea and argued that *history was at the top of the priority list of all knowledge* that is conducive to human development. He proposed a three-phase theory on the development of history: phase one was theology; phase two, metaphysics; and phase three, positivism—which is supported by rational science. The course of history would achieve an optimum stage, that is, the right foundation of social progress. The founder of history of science, George Sarton, was a champion of the positivism-based social progress theory.

If we are to borrow the theories of the German philosopher and historian, Heinich Rickert (1863-1936), the historical researches conducted before and by George Sarton were, in fact and in general, individualized knowledge (such as Newton's manuscripts on optics and Galileo's experiment in mechanics), guided by a value system (for instance, the social progress theory). Conversely, historians who analyze science holistically focus their research on generalized and disciplinary conclusions, without being influenced by any certain value system. Thus, born out of rigorous historical studies, history of science—thanks to external evolution and internal adjustment of its research orientation— has in it characteristic features that help to assimilate it to that ascribed by the natural sciences as their objectives [Rickert, 1962]. Such a knowledge system comes closest to the goals of harmonizing the two cultures.

If we may say that Sarton failed to realize his new humanitarian creed, then in the 21$^{st}$ century, with the dawning of globalization, it is possible that we may integrate scientific and humanistic cultures using history of science.

# References

Cho, Co-ching (Zhu, Kezhen) [1926] "Climatic pulsations during historical times in China," *Geographical Review* **16**(2), 274-283.

Cho, Ko-chen. (Zhu, Kezhen) [1973] "A preliminary study on the climatic fluctuations during the last 5000 years in China," *Scientia Sinica* **16**(2), 226-256.

Hsi, Tse-Tsung (Xi, Zezong) [1955] "A new catalogue of ancient novae," *Acta Astronomica Sinica* **2**, 183-196. English translation in *Smithsonian Contributions to Astrophysics* **2**(6).

IUHSP/DHST [2005] "Beijing Declaration," The 22nd International Congress of History of Science, July 24-30, Beijing, China.

Lam, L. [2008] "Science Matters: A unified perspective," in *Science Matters: Humanities as Complex Systems*, eds. Burguete, M. & Lam, L. (World Scientific, Singapore).

Li, Y. & Qian, B.C. [1998] *Corpora of Li Yan and Qian Baocong on History of Science*, 10 vols., eds. Guo, S.C. & Liu, D. (Liaoning Education, Shenyang).

Liu, D. [1999a] "The cultural function of history of science and its institutionalization," *Journal of Dialectics of Nature* **21**(3), 75-76.

Liu, D. [1999b] "History of science towards the 21st century," *Studies in the History of Natural Sciences* **18**(3), 193-195.

Liu, D. [2000a] "History of science, scientific strategy, and innovative culture," *Journal of Dialectics of Nature* **22**(1), 4-6.

Liu, D. [2000b] "A new survey of the Needham Question," *Studies in the History of Natural Sciences* **18**(4), 293-305.

Liu, D. [2006] *Culture ABC* (Hubei Education Press, Wuhan) pp. 57-63.

Needham, J. [1954] *Science and Civilisation in China*, Vol. 1 (Cambridge University Press, Cambridge, UK) pp. 3-4.

Needham, J. [1967] "The roles of Europe and China in the revolution of oecumenical science," *Journal of Asian History* **1**(1), 3-32.

Rickert, H. [1962] *Science and History: A Critique of Positivist Epistemology*, English translated by Reisman, G. (van Nostrand, London).

# PART III

# Raising Scientific Level

# 11

# Why Markets Are Moral

*Michael Shermer*

The new sciences of evolutionary economics, behavioral economics, and neuroeconomics demonstrate how and why markets must be moral in order to function. Trust is the key to understanding market exchange. This chapter discusses the relationship between trust and economic prosperity, as well as the neurochemistry of trust in the form of oxytocin, a hormone that increases trust between strangers in an economic exchange game. Implications for trust, trade, and economic prosperity are considered from this new research.

## 11.1 The Neurochemisty of Trust

"There's an old English proverb that says *It is an equal failing to trust everyone and to trust no one*." So begins Paul Zak, a professor of economics at Claremont Graduate University who is taking the study of economic behavior down to the molecular level in his search for the neurochemistry of trust and trade, which he believes is grounded in oxytocin, a hormone synthesized in the hypothalamus and secreted into the blood by the pituitary. In women, oxytocin stimulates birth contractions, lactation, and maternal bonding with a nursing infant. In both women and men it increases during sex and surges at orgasm, playing a role in pair bonding, an evolutionary adaptation for long-term care of helpless infants. "We know that trust is a very strong predictor of national prosperity, but I want to know what makes two people trust one another," Zak explains as we sit down in his Center for Neuroeconomics

Studies nestled in the bedroom community of Claremont, California.[1]

Zak is the oxytocin man. It says so right on his license plate. Tall and handsome with square shoulders and the physique of someone who works out regularly, Zak's firm grip and warm smile exude, well, trust. Trained in traditional economics, in the mid 1990s his research led him to connect trust to economic growth. A 2001 study on trust in forty-two countries, for example, asked people in their native language, "Generally speaking, would you say that most people can be trusted, or that you cannot be too careful in dealing with people?" The results were as diverse as they were striking. At the low end of the trust scale, only 3 percent of those surveyed in Brazil and 5 percent in Peru believe that their fellow citizens are trustworthy, compared to 65 percent of Norwegians and 60 percent of Swedes who trust one another. Falling in the middle of the scale were the United States, at 36 percent, and the United Kingdom, at 44 percent. The rankings remain essentially unchanged even when they are controlled for income. Trust is high in the countries of Scandinavia and East Asia but low in the countries of South America, Africa, and especially in the former Communist bloc. "The simple correlation between national rates of investment (gross investment per Gross Domestic Product) and trust is strongly positive," Zak continues, so that "when trust is low, investment lags. The same positive correlation holds for GDP growth and trust."

Economic mechanics drive the relationship between trust and prosperity. "Trust facilitates transactions by reducing the number of contingencies that must be considered when 'doing a deal.' A deal sealed with a handshake between principals can only occur in a high-trust situation. Let the lawyers work out the details—we have a deal," Zak offers. "Conversely, when trust is low, negotiations are protracted, and therefore more costly. When transaction costs are higher, fewer transactions occur and investment and economic growth are lower. Trust is among the most powerful stimulants for investment and economic growth that economists have discovered. In seeking to understand why some countries are poor and others are rich, it is, therefore, crucial to

---

[1] Interviews conducted over a series of visits in February and March of 2007.

understand the foundation for interpersonal trust." Perhaps this is why countries that have higher rates of generalized trust show higher rates of return on national stock markets [Zak, 2003]. From these and other studies, Zak realized that in order for a nation to achieve prosperity it is vital to maximize positive social interactions among its members in order to increase trust.

The list of positive social interactions identified by Zak's research will surprise no one living in a liberal democracy with relatively free markets: protection of civil liberties, freedom of the press, freedom of association, freedom of travel (good roads and reliable infrastructure), freedom of communication (working phone systems), mass education, a reliable banking system, a sound currency, and especially the freedom to trade [Zak & Knack, 2001]. He even found a connection between a clean environment and trust, whereby people in countries with polluted environments show higher levels of estrogen antagonists, which lower their levels of oxytocin—and thus their feelings of trust. Zak went so far as to compute the differences in living standards that trust can effect, whereby "a 15 percent increase in the proportion of people in a country who think others are trustworthy raises income per person by 1 percent per year for every year thereafter." For example, increasing levels of trust in the United States from its present 36 percent to 51 percent would raise the average income for every man, woman, and child in the country by $400 per year, or $30,000 over a lifetime [Zak, 2003]. Trust has fiscal benefits.

## 11.2  Gaming the Market

The connection between social interactions and trust can also be seen in the laboratory, when subjects participate in experiments utilizing the Prisoner's Dilemma game paradigm. In the scenario, each of two subjects pretends to be a prisoner arrested for a crime. Each is independently made the same offer, and both assume the other has been presented the same deal. Imagine that you are the subject and these are your options:

1.  If you and your partner both cooperate then you each get one year in jail.
2.  If you confess that both you and your partner committed the crime, then you go free and your partner gets three years.
3.  If your partner confesses and you don't, then you receive the three-year penalty while he goes free.
4.  If you both confess then you each get two years.

If you defect on your partner and confess, then you will either get 0 or 2 years, depending on what he does. If you cooperate and stay quiet, you either get 1 or 3 years, again depending on his response. The logical choice, then, is to defect. Of course, your partner is likely going to make the same calculation as you, which means he too will defect, guaranteeing that you will receive a two-year punishment. However, you also then come to the conclusion that he is probably computing this same strategy, and hope that as a result he will realize that you should both cooperate. But then, if that is his line of reasoning, perhaps he'll defect in hopes that you cooperate assuming that he cooperates, getting him off the hook while you get hit with the three-year penalty. This is why it is called a dilemma.

When the game is played just once, both people typically defect. But when the game is played over a fixed number of rounds, defection is the norm, because each person realizes that the other player will defect on the last round, and thus is spurred to defect in the second to the last round, which spurs the other player to defect a round earlier, and so forth, until both players are defecting from the first round. However, when the game is played in something more similar to the real world with the game played over an unknown number of rounds—as would be the case in a negotiation process, where the number of steps is typically unknown—both players keep track of what the other has been doing round by round, and cooperation prevails. Over and over it has been shown that the best strategy in a series of Prisoner's Dilemma games without a known end point is "tit-for-tat" with no initial defection. That is, the most self-benefiting thing to do in the long run is to begin by trusting and cooperating and then do whatever your partner does. In such a contest, the winning strategy is called "Firm but Fair," and calls for you to cooperate with cooperators, to cooperate after a mutual defection, to quit playing with constant defectors, and to defect with partners who always cooperate (otherwise known as suckers) [Hutson & Vickers, 1995; Binmore, 1994]. Even more realistic simulations include the "Many Person Dilemma," in which one player interacts with several other players, and in conditions where subjects are allowed to accumulate experience with the other players thereby giving them an opportunity to establish trust. Finally, the Prisoner's Dilemma protocol has been applied to the real world of business negotiations, marital disputes, and cold war strategies. It turns out that in both computer simulations and in the real-world, cooperation is by far the dominant strategy.[2]

In a related experiment on cooperation, nine subjects were each given \$5. If five or more of the nine cooperated by donating their \$5 to a general pot, all nine would receive \$10. Although it pays to be a cooperator (you get \$10 instead of \$5), it pays even more to be a defector (\$15 instead of \$5), as long as at least five other people cooperate. The results were mixed, with many groups of nine subjects failing to achieve the critical mass of five cooperators, because there was no cooperation. Then the experimenters added a step: members of some groups were given the opportunity to discuss their strategy options before playing. Those groups that interacted before playing averaged eight cooperators, and 100 percent of these groups earned cooperative bonuses. By sharp contrast, those groups that did not interact before playing earned bonuses only 60 percent of the time [Bower, 1990]. In a similar experiment on social dilemmas, psychologist Robyn Dawes found that groups given the opportunity to communicate face to face were more likely to cooperate than those who were not. "It is not just the successful group that prevails," Dawes concluded, "but the individuals who have a propensity to form such groups" [Dawes *et al.*, 1990].

Where in the brain are these dilemmas resolved? Employing the

---

[2] There is a sizable body of literature on game theory and cooperation. See for example: [Trivers, 1971; Axelrod & Hamilton; 1981; Axelrod, 1984; Hofstadter, 1983]. On the Internet: google search "prisoner's dilemma" will lead to thousands of sites, computer simulations, chat rooms, discussions, bibliographies, and so on, such as http://pespmcl.vub.ac.be/PRISDIL.html. See also: [Frank, 1988; Ridley, 1996; Murnighan, 1992; Taylor, 1987].

Prisoner's Dilemma protocol while scanning thirty-six subjects' brains using an fMRI, James Rilling and his colleagues at Emory University found that in cooperators the brain areas that lit up were the same regions activated in response to such stimuli as desserts, money, cocaine, and beautiful faces. Specifically, the neurons most responsive were those rich in dopamine located in the anteroventral striatum in the middle of the brain—the so-called "pleasure center." Tellingly, cooperative subjects reported increased feelings of trust toward and camaraderie with like-minded partners [Rilling *et al.*, 2002].

These findings make sense in an evolutionary model, where informal and non-codified rules of conduct developed within small bands of hunter-gatherers because knowing all the other players in the game leads to the evolution of cooperation [Forsythe *et al.*, 1994]. The psychological impulse to form relationships and alliances is the deeper cause that lies beneath the moral sentiment of trust, and trade is an effective medium that allows people to create trusting relationships with and form attachments to other trustworthy people. In other words, it is not enough to fake being a cooperator, because over time and with experience deceivers are usually flushed out. You actually have to believe you are a cooperator, and there is no surer way to believe you are a cooperator than to actually be one, believe it, and mean it. As Yogi Berra counseled, "Always go to other people's funerals; otherwise, they won't go to yours." From behavior to brains to blood, the molecular basis of trust begins at the most fundamental level of human relations. In her book *Why We Love*, anthropologist Helen Fisher makes a distinction between lust and love. Lust, she says, is enhanced by dopamine, the neurohormone produced by the hypothalamus that is associated with reward and pleasure. Fisher shows that it also triggers the release of testosterone, the hormone that drives sexual desire, and thus dopamine is implicated in yet another vital component of human relations—the yearning desire for another [Fisher, 2004]. Love, by contrast, is the emotion of attachment reinforced by oxytocin.

## 11.3  Trust and Trade

Here we return to the connection between trust, trade, and oxytocin, and Paul Zak's theory that there is a direct connection between the three. "Oxytocin and testosterone are two branches of this trust-distrust system, and as we go through the world we are constantly balancing these levels of trust and distrust," Zak explained as he recounted some of the experiments he and his team have been running in his center. In one experiment, for example, Zak found that oxytocin increases "when a person observes that someone wants to trust him or her." It turns out, in fact, that with the exception of the two percent of the population who are sociopathic, when someone trusts, it triggers the release of oxytocin. Although this is the case for both men and women, Zak has found that "when women get a boost of oxytocin they are more likely to reciprocate than men are. Men are more sensitive to violations of social norms, and when the norms are broken men get a disproportional rise in testosterone. Women don't like being distrusted, but they don't have this heated and angry response that men get, which is related to testosterone."

Oxytocin is deeply involved in the attachment process, whereas testosterone is intensely caught up in the enforcement of the social norms (which may help to explain why far more men go into such professions as the military and law enforcement). In exchange games, the more subjects are behaving in trusting ways, the more money they exchange and the higher the levels of oxytocin that are released by the brain [Zak *et al.*, 2004; Zak *et al.*, 2005]. When Zak asks them *why* they are giving up so much money, subjects say such things as "it just feels right." The moral emotion drives behavior, even if the moral calculation beneath the emotion is invisible.

Skeptics might reasonably ask whether oxytocin is the cause or the effect of trust. "To control for that," Zak says, "we set up an experimental condition where instead of subjects freely choosing to trust someone, we had them randomly pull out a ping pong ball from a box that would determine how much money was given or received, and in those cases there was no significant change in oxytocin levels." To find out if cooperating and trust lead to the release of oxytocin or if increased levels of oxytocin lead to more cooperation and trust, Zak infused oxytocin into subjects' brains through a nose spray that is quickly absorbed by the body and discovered that it causes them to act more cooperatively.

Zak has surmounted a number of additional lines of evidence to support his thesis. *Trust and happiness*: People who trust and are trustworthy report being happier. *Trust and touch*: We all know how good it feels to be touched by someone, so Zak ran an experiment in which he gave subjects in an exchange game a massage by a licensed massage therapist, which led those who received the massage and received a signal of trust to offer up to 250 percent more than subjects who did not receive the massage and trust signal. *Trust and smell*: Oxytocin may also be mediated by smell, Zak suggests, noting that there are oxytocin receptors on the olfactory bulb and citing an experiment in which the smell of a mother's own newborn baby triggers the release of oxytocin and incredibly strong feelings of attachment. *Trust and neglect*: Animals that are abused or neglected shortly after birth show a loss of regions of the brain that have oxytocin receptors, and those animals become withdrawn, socially inappropriate, and depressed [Zak, 2005; Kosfeld *et al.*, 2005; Morhenn *et al.*, 2008]. The implications of Zak's research are profound. "Oxytocin is the social glue of society. It is what keeps us together as a civilization. If we didn't have something in our head that indicated who we should trust and who we should not, civilization wouldn't work. We're social animals and we need a trust detector in our heads."

Zak's new findings about oxytocin get at something very deep in the evolutionary origins of morality: the role of evil people in society. In my book *The Science of Good and Evil* [Shermer, 2004], I attributed evil to our dual dispositional nature and the fact that in addition to being trusting, cooperative, and altruistic, we are also distrusting, competitive, and selfish, and that the evolutionary forces that led us to be pro-social with our fellow in-group members also led us to be tribal and xenophobic against out-group strangers. Zak pushes the evolutionary model further by looking not just at the potential for evil that resides in all of us, but at the anomaly of evil that lurks within the two percent of individuals who are sociopaths. While sociopaths comprise three to four percent of the male population and less than one percent of the female population, they are believed to account for 20 percent of the U.S. prison population and between 33 percent and 80 percent of the chronic criminal offender population. Altogether, these individuals may account for half of all crimes in the United States [Mealy, 1995; Lykken, 1995].

In his social game experiments, Zak typically finds that about two percent of his subjects (he calls them "bastards") do not respond to oxytocin or other social cues that normally encourage trust and cooperation. "These individuals have a dysfunction in their oxytocin release and have an identifiable difficulty attaching to others." But instead of seeing this as nothing more than a biological accident in the wiring of the brain, Zak thinks there may be adaptive evolutionary reasons behind such misfirings. "Bastards are necessary from an evolutionary standpoint because they keep the physiologic balance between appropriate levels of trust and distrust optimally tuned. Without these exceedingly selfish people, humans might have evolved into being unconditional trusters. If so, we would become susceptible to invasion by those who would prey on our perfectly trusting nature."

As part of his evidence, Zak cites the case of a woman who has a rare genetic disorder that caused her amygdala to calcify and die. "The amgydala is a primary target for oxytocin in the brain that helps maintain the trust/distrust balance," he notes, "and she has difficulty reading the trustworthiness of faces, is very impulsive in her decision-making, and is terrible with money. She has normal I.Q., but is often a target for unscrupulous predators who sense her unconditional trust in others and take money from her." The Caltech neuroscientist Ralph Adolphs examined the woman for other deficiencies and discovered that she is unable to recognize fear in the face of others. When you look at another person's face, your eyes rapidly dart about, scanning for clues and collecting data about its emotional expression, which, according to the psychologist Paul Ekman and many others, evolved to be universal and thus identifiable. The amygdala-damaged woman, by contrast, stares straight at the face without scanning for details, and thus is unable to make an emotional assessment [Adolphs *et al.*, 1998; Adolphs *et al.*, 1994]. During social interactions, most of us maintain a balance between trust (mediated by oxytocin) and fear (mediated by epinephrenie, norephinepherine, and other stress hormones), and we quickly and unconsciously adapt to our environment and the people in it. But when that balance is broken, so too is our trust detector [Zak, 2008].

## 11.4  The Evolution of Trust and Trade

Zak's model fits well with my own that we evolved an innate sense of right and wrong that is expressed through the moral emotions, and that free trade is an integral component to breaking down the normal tribal barriers blocking trust. That being the case, economic transactions can occur with a minimum of top-down interference. A *shadow of enforcement*—a hint of potential punishment for a norm violation—is all that is needed for most people in most circumstances to grease the wheels of commerce. A law enforcement dummy placed along a stretch of highway, for instance, has been shown to get motorists to slow down, not because they thought it was an actual enforcer but because it reminded them of the law.

Trade makes people more trusting and trustworthy, which makes them more inclined to trade, which increases trust... creating a self-enforcing cycle of trust, trade, freedom, and prosperity. An additional evolutionary connection between trust and trade can be seen in a two-part experiment on cooperation and cheating conducted by Dan Chiappe and his colleagues at the California State University, Long Beach, in which subjects first categorized individuals as cheaters, cooperators, or neither based on a written description of how they behaved in an exchange involving, for example, borrowing and paying back (or not) money. After reading the description, each subject was then asked to rate how important it was to remember the individuals using a 7-point scale. In the second experiment, participants categorized the individuals on the 7-point scale as before, but then had the opportunity to look at photographs of their faces. They then participated in a face recognition test. The first experiment found that cheaters were rated more important to remember than cooperators—especially when a greater amount of resources was involved—and cooperators were rated more important to remember than those categorized as neutral. The second experiment showed that cheaters were looked at longer and the subjects had a better memory for their faces [Chiappe *et al.*, 2004].

Why should this be? Because from an evolutionary economics perspective, cheaters—like the two percent of individuals who routinely break the laws of society—keep cooperators on their moral toes. It is therefore vital to be able to discriminate between the cheaters and the cooperators. As Chiappe explained: "Everything else being equal, knowing that a person cooperated may not tell us much about their character. Knowing that they cheated would be more relevant. This is because cheaters have to give the appearance of being trustworthy and thus they may have to cooperate much of the time." Indeed, most people most of the time in most circumstances are cooperative so it takes a lot more mental data storage space to focus on and remember details about them (such as their faces) compared to the small handful of cheaters, whose social norm violations stand out as the exception and thus must be recalled for future transactions. And as research in memory has consistently demonstrated, we are more likely to recall unusual events than we are common occurrences. The findings also indicate that we remember cooperators more than neutral people. This too makes sense from an evolutionary perspective, since the memory gradient from cheaters to cooperators to neutrals would be adaptive in our management of social relations in all forms of human exchange, including and especially economic trade.

If this theory is true, there should be neural networks associated with cooperation in an exchange task, and sure enough, the George Mason University neuroeconomist Kevin McCabe and his colleagues, in a fMRI brain scan study of subjects who participated in a "trust and reciprocity" game with either another person or a computer, found that players who were more trusting and cooperative showed more brain activity in Brodmann area 10, associated with reading other's intentions, and more activity in the limbic system, which is associated with the processing of emotions. Tellingly, such activity was not seen in this brain region of players who either played against computers or engaged in non-cooperative games, which makes perfect sense in a social primate species that evolved moral emotions that drive such pro-social behaviors [McCabe *et al.*, 2001].

## 11.5  The Evolution of Fairness, or Why We are Moral

Evolutionary economics suggests that instead of thinking of human culture as being cleaved by different national economic institutions and varying individualized personal values, we should think of it as an evolved human nature that gives rise to a set of core institutions and values that vary in details across different cultures.

As a test of this interpretation, over the past quarter century hundreds of experiments in behavioral economics have been conducted on people from dozens of countries around the world, including fifteen small-scale indigenous tribes. Take the Ultimatum Game, where one player proposes how to divide a sum of money and the second player accepts or rejects the proposal. For example, say you are given $100 to split between yourself and your game partner. Whatever division of the money you propose, if your partner accepts it, you are both richer by that amount. How much should you offer? Why not suggest a $90-$10 split? If your game partner is a rational self-interested money-maximizer—as predicted by the standard economic model of *Homo economicus*—he isn't going to turn down a free ten bucks, is he? He is. Research shows that proposals that deviate much beyond a $70-$30 split are usually rejected.

Why? Because they aren't fair. Says who? Says the moral emotion of "reciprocal altruism," which evolved over the Paleolithic eons to demand fairness on the part of our potential exchange partners. "I'll scratch your back if you'll scratch mine" only works if I know you will respond with something approaching parity. The moral sense of fairness is hardwired into our brains and is an emotion shared by all people and primates tested for it. Thousands of experimental trials have consistently revealed a sense of injustice at low-ball offers. This includes peoples in non-Western cultures around the world, including those living close to how our Paleolithic ancestors lived, and although their responses vary more than modern peoples living in market economies do, they still show a strong aversion to unfairness [Henrich *et al.*, 2001].

Any theory of economics must begin with a sound theory of human nature. Evolutionary economics redefines the borders of our nature, showing just how driven we are by ancient programs designed for another time and another place. But we also evolved the *adaptation of adaptability*, and it is here where we see how and why humans behave as we do in such social institutions as markets: we cooperate for the same reason we copulate—because it feels good. On a deeper evolutionary level, the reason cooperating feels good is because it is good for us, individually and as a species. Trust and cooperation leads to a viable free market of exchange, and free markets lead to greater trust and cooperation [Shermer, 2007].

# References

Adolphs, R., Tranel, D., Damasio, H. & Damasio, A. R. [1994] "Impaired recognition of emotion in facial expressions following bilateral damage to the human amygdale," *Nature* **372**, 669-672.

Adolphs, R., Tranel, D. & Damasio, A. R. [1998] "The human amygdale in social judgment," *Nature* **393**, 470-474.

Axelrod, R. & Hamilton, W. D. [1981] "The evolution of cooperation," *Science* **211**, 1390-1496.

Axelrod, R. [1984] *The Evolution of Cooperation* (Basic Books, New York).

Binmore, K. [1994] *Game Theory and the Social Contract. Vol. 1: Playing Fair* (MIT Press, Cambridge, MA).

Bower, B. [1990] "Getting out from number one: Selfishness may not dominate human behavior," *Science News* **137**(17), 266-267.

Chiappe, D., Brown, A., Dow, B. Koontz, J., Rodriguez, M. & McCulloch, K. [2004] "Cheaters are looked at longer and remembered better than cooperators in social exchange situations," *Evolutionary Psychology* **2**, 108-120.

Dawes, R. M., van de Kragt, A. & Orbell, J. M. [1990] "Cooperation for the benefit of us—not me, or my conscience," in *Beyond Self-Interest*, ed. Mansbridge, J. (University of Chicago Press, Chicago) pp. 97-110.

Fisher, H. [2004] *Why We Love: The Nature and Chemistry of Romantic Love* (Henry Holt, New York).

Forsythe, R., Horowitz, J. L., Savin, N. E. & Sefton, M. [1994] "Fairness in simple bargaining experiments," *Game and Economic Behavior* **6**, 347-369.

Frank, R. [1988] *Passions within Reason: The Strategic Role of the Emotion* (W.W. Norton, New York).

Henrich, J., Boyd, R., Bowles, S., Camerer, C. Gintis, H., McElreath, R. & Fehr, E. [2001] "In search of *Homo economicus*: Experiments in 15 small-scale societies," *American Economic Review* **91**(2), 73-79.

Hofstadter, D. R. [1983] "Metamagical themes: Computer tournaments of the Prisoner's Dilemma suggest how cooperation evolves," *Scientific American*

248(5), 16-26.

Hutson, V. C. L. & Vickers, G. T. [1995] "The spatial struggle of tit-for-tat and defect," *Philosophical Transactions of the Royal Society of London B* **348**, 393-404.

Kosfeld, M., Heinrichs, M., Zak, P. J., Fischbacher, U. & Fehr, E. [2005] "Oxytocin increases trust in humans," *Nature* **435**, 673-676.

Lykken, D. T. [1995] *The Antisocial Personalities* (Lawrence Earlbaum Associates, Hillsdale, NJ).

McCabe, K., Houser, D., Ryan, L., Smith, V. & Trouard, T. [2001] "A functional imaging study of cooperation in two-person reciprocal exchange," *Proceedings of the National Academy of Sciences, USA* **98**, 11832-11835.

Mealy, L. [1995] "The sociobiology of sociopathy," *Behavioral and Brain Sciences* **18**, 523-599.

Morhenn, V. B., Park, J. W., Piper, E. & Zak, P. J. [2008] "Monetary sacrifice among strangers is mediated by endogenous oxytocin release after physical contact," *Proceedings of the National Academy of Science, USA* (in press).

Murnighan, J. K. [1992] *Bargaining Games* (William Morrow and Co., New York).

Ridley, M. [1996] *The Origins of Virtue* (Viking, New York).

Rilling, J., Gutman, D. A., Zeh, T. R., Pagnoni, G., Berns, G. S. & Kilts, C. D. [2002] "A neural basis of social cooperation," *Neuron*, July 18, **35**, 395-404.

Shermer, M. [2004] *The Science of Good and Evil: Why People Cheat, Gossip, Care, Share, and Follow the Golden Rule* (Henry Holt/Times Books, New York).

Shermer, M. [2007] *The Mind of the Market: Compassionate Apes, Competitive Humans, and Other Tales from Evolutionary Economics* (Henry Holt/Times Books, New York).

Trivers, R. L. [1971] "The evolution of reciprocal altruism," *Quarterly Review of Biology* **46**, 35-57.

Zak, P. J. & Knack, S. [2001] "Trust and growth," *The Economic Journal* **111**, 295-321.

Zak, P. J. [2003] "Trust," *Capco Institute Journal of Fiancial Transformation* **7**, 13-21.

Zak, P. J., Kurzban, R. & Matzner, W. T. [2004] "The neurobiology of trust," *Annals of the New York Academy of Sciences* **1032**, 224-227.

Zak, P. J. [2005] "Trust: A temporary human attachment facilitated by oxytocin," *Behavioral and Brain Sciences* **28**(3), 368-369.

Zak, P. J., Kurzban, R. & Matzner, W. T. [2005] "Oxytocin is associated with human trustworthiness," *Hormones and Behavior* **48**, 522-527.

Zak, P. J. [2008] "Values and value: Moral economics," in *Moral Markets: The Critical Role of Values in the Economy*, ed. Zak P. J. (Princeton University Press, Princeton).

# 12

# Towards the Understanding of Human Dynamics

*Tao Zhou, Xiao-Pu Han and Bing-Hong Wang*

Quantitative understanding of human behaviors provides elementary but important comprehension of the complexity of many human-initiated systems. A basic assumption embedded in previous analyses on human dynamics is that its temporal statistics are uniform and stationary, which can be properly described by a Poisson process. Accordingly, the interevent time distribution should have an exponential tail. However, recently, this assumption is challenged by extensive evidence, ranging from communication to entertainment to work patterns, that human dynamics obeys non-Poisson statistics with heavy-tailed interevent time distribution. This chapter reviews and summarizes recent empirical explorations on human activity pattern, as well as the corresponding theoretical models for both task-driven and interest-driven systems. Finally, we outline some open questions in the studies of statistical mechanics of human dynamics.

## 12.1 Introduction

Human behavior, as an academic issue in science, has a history of about one century since the time of Watson [1913]. As a joint interest of sociology, psychology and economics, human behavior has been extensively investigated during the last decades. However, due to the complexity and diversity of our behaviors, the in-depth understanding of human activities is still a long-standing challenge thus far. Actually, up to now, most of academic reports on human behaviors are based on clinical records and laboratorial data, and most of the corresponding hypotheses and conclusions are only qualitative in nature. Therefore, we have to ask at least two questions:

1. Could those laboratorial observations properly reflect the real-life activity pattern of us?
2. Can we establish a quantitative theory for human behaviors?

Barabási [2005] provided a potential starting point in answering those questions, which is to extract the statistical laws of human behaviors from the historical records of human actions. Traditionally, the individual activity pattern is usually simplified as a completely random point-process, [1] which can be well described by a Poisson process, [2] leading to an exponential interevent time[3] distribution [Haight, 1967]. That is, the distribution of time difference between two consecutive events should be almost uniform, and a long gap is hardly to be observed. However, recently, both empirical studies and theoretical analyses display a far different scenario: our activity patterns follow non-Poisson statistics, characterized by bursts of rapidly occurring events separated by long gaps. These new findings have significant scientific and commercial potential. As pointed out by Barabási [2005], models of human activities are crucial for better resource allocation and pricing plans for telephone companies, to improve inventory and service allocation in both online and "high street" retail.

Barabási and his colleagues have opened up a new research area, namely, *Human Dynamics*. Still in its infancy but motivated by both theoretical and practical significances, the studies of human dynamics attract more and more attention. In this chapter, we summarize recent progresses on this topic, which may be helpful for the comprehensive understanding of the architecture of complexity [Barabási, 2007]. This chapter is organized as follows. In the next section, we show the empirical evidence of non-Poisson statistics of human dynamics. In Sections 12.3 and 12.4, the task-driven and interest-driven models are

---

[1] A completely random point-process is one such that at each discrete time step, an event happens with a constant probability, the magnitude of which is independent of time.
[2] A Poisson process is mathematically defined by Eq. (12.1), or equivalently Eq. (12.2), below.
[3] Interevent time is the time duration between two consecutive events of the same nature, to be specified in each concrete case under discussion. See examples in Section 12.2.

introduced. Finally, we outline some open problems in studying the statistical mechanics of human dynamics.

## 12.2  Non-Poisson Statistics of Human Dynamics

Previously, it is supposed that temporal statistics of human activities can be described by a Poisson process. In other words, for a sufficiently small time difference $\Delta t$, the probability that one event (i.e., one action) occurs in the interval $[t, t+\Delta t)$ is independent of time $t$, and has approximately a linear correlation with $\Delta t$ such as

$$P(t, t+\Delta t) \approx \lambda \Delta t, \quad \lambda > 0, \tag{12.1}$$

where $\lambda$ is a constant. Under this assumption, the interevent time between two consecutive events, denoted by $\tau$, obeys an exponential distribution [Haight, 1967]:

$$P(\tau) = \lambda e^{-\lambda \tau}. \tag{12.2}$$

Equation (12.2) represents an exponential decay, implying that a long gap without any event (i.e., a very large $\tau$) should be rarely observed. However, we will show below extensive empirical evidence in real human-initiated systems, where the distributions of interevent time have much heavier tails than the one predicted by Poisson process. Hereinafter, we will review the empirical results for different systems one at a time.

## 1.  Email communication

The data set contains the email exchange between individuals in a university environment for three months [Eckmann *et al.*, 2004]. There are in total 3,188 users and 129,135 emails with second resolution. Denote by $\tau$ the interevent time between two consecutive emails sent by the same user, and $\tau_w$ the response time taking for a user to reply a received email. As shown in Fig. 12.1, both the distributions of interevent time and response time obey a power law[4] with exponent approximately equal to -1. Although the exponent differs slightly from user to user, it is typically centered on -1.

## 2. Surface-mail communication

The data set used for analysis contains the correspondence records of three great scientists: Einstein, Darwin and Freud. The sent/received numbers of letters for those three scientists are 14,512/16,289 (Einstein), 7,591/6,530 (Darwin), and 3,183/2,675 (Freud). The dataset is naturally incomplete, as not all letters written or received by these scientists were preserved. Yet, assuming that letters are lost at a uniform rate, they should not affect the main statistical characteristics. As shown in Fig. 12.2 [Oliveira & Barabási, 2005; Vázquez *et al.*, 2006], the distributions of response time follow a power law with exponent approximated to -1.5. The readers should be warned that the power-law fitting for *Freud* is not as good as that for *Einstein* or *Darwin*. Recently, we analyze the correspondence pattern of a Chinese scientist, *Xuesen Qian*, and find that both the distributions of interevent time and response time follow a power law with exponent -2.1 [Li *et al.*, 2007].

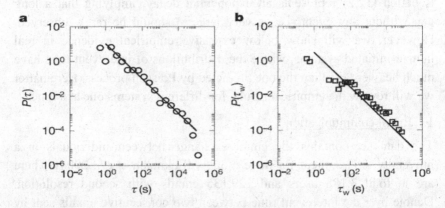

Fig. 12.1. Heavy-tailed activity patterns in email communications [Barabási, 2005]. (a) Distribution of interevent time; (b) distribution of response time. Data shown in these two plots are extracted from one user. Both solid lines have slope -1 in the log-log plot.

---

[4] A power laws means two variables $y$ and $x$ are related by the relationship, $y \propto x^a$, where the exponent $a$ is a constant.

## 3. Short-message communication

The data set contains the records of short-message communications of a few volunteers using cell phones in China [Hong *et al.*, 2008]. Figure 12.3 shows two typical distributions of interevent time between two consecutive short messages sent by an individual. Both distributions can be well fitted by a power law with different exponents. Actually, in the data set, almost all distributions of interevent time can be well approximated by a power law, and there is apparently correlation between the average numbers of short messages sent out per day and the power law exponents.

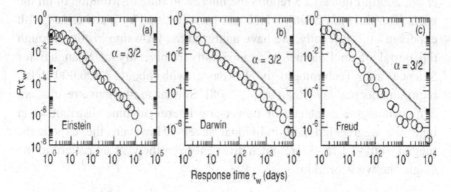

Fig. 12.2. Distributions of response times for letters replied by Einstein, Darwin and Freud, respectively [Vázquez *et al.*, 2006]. All three distributions are approximated by a power law tail with exponent -1.5.

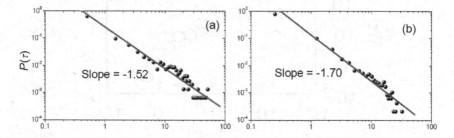

Fig. 12.3. Typical examples of distributions of time interval of sending short-messages on log-log plots [Hong *et al.*, 2008]. The x axis denotes time interval (in hour); y axis, the probability. The two distributions are approximated by a power law with exponent -1.52 and -1.70, respectively.

## 4.  Web browing

Web browsing history can be automatically recorded by setting cookies. The data set contains the visiting records of 250,000 visitors to the site www.origi.hu from Nov. 8 to Dec. 8 in the year 2002, with about 6,500,000 HTML hits per day [Dezsö *et al.*, 2006]. Figure 12.4 shows the interevent time distribution of a single user, which can be approximately fitted by a power law with exponent -1.0. Although the power law exponents for different users are slightly different, they centered around -1.1. Actually, the exponent distribution obeys a Guassian function with characteristic value approximate to 1.1 [Vázquez *et al.*, 2006]. Figure 12.5 reports the interevent time distribution of all the users [Dezsö *et al.*, 2006], which can be well fitted by a power law with exponent -1.2. Recently, we have analyzed the browsing history through the portal of an Ethernet of a university (University of Shanghai for Science and Technology) in 15 days, with about 4,500,000 URL requirements per day [Zhao *et al.*, 2008]. Similar to the above results, we demonstrate the existence of power-law interevent time distribution in both the aggregated and individual levels. However, the exponents, ranged from -2.1 to -3, are much different from the visiting pattern to the single site www.origi.hu.

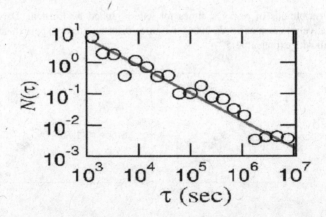

Fig. 12.4. The distribution of interevent time between two consecutive web visits by a single user [Vázquez *et al.*, 2006]. $N(\tau)$ stands for the frequency. The solid line has slope -1.0.

## 5. Library loan

The data set contains the time books or periodicals were checked out from the library by the faculty at the University of Notre Dame during three years [Vázquez *et al.*, 2006]. The number of borrowers is 2,247, and the total transaction number is 48,409. The interevent time is the time difference between consecutive books or periodicals checked out by the same patron. Figure 12.6 presents the interevent time distribution of a typical user, which can be well fitted by a power law with exponent about -1.0. The exponents for different users are different, ranging from -0.5 to -1.5, with the average around -1.0.

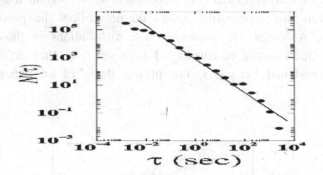

Fig. 12.5. Distribution of interevent time between two consecutive web visits by all users [Dezsö *et al.*, 2006]. $N(\tau)$ stands for the frequency. The solid line has slope -1.2.

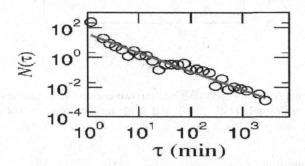

Fig.12.6. Distribution of interevent time between two consecutive books or periodicals checked out by the same user [Vázquez *et al.*, 2006]. $N(\tau)$ stands for the frequency. The solid line has slope -1.0.

## 6.  Financial activities

The data set contains all buy/sell transactions initiated by a stock broker at a Central European bank between June 1999 and May 2003, with an average of ten transactions per day and a total of 54,374 transactions [Vázquez *et al.*, 2006]. The interevent time represents the time between two consecutive transactions by the broker, with the gap between the last transaction at the end of one day and the first transaction at the beginning of the next trading day ignored. Different from the empirical systems above, the interevent time distribution for stock transactions obviously departs from a power law (Fig. 12.7). A recent empirical analysis on a double-auction market [Scalas *et al.*, 2006] also shows that the interevent time between two consecutive orders do not follow the power-law distribution. Although the interevent time distributions in these two financial systems cannot be considered as power laws, they do display clearly heavy-tailed behaviors, not in the shape of an exponential function.

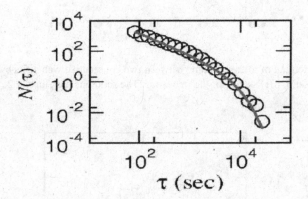

Fig. 12.7. Distribution of interevent time between two consecutive transactions initiated by a stock broker [Vázquez *et al.*, 2006]. $N(\tau)$ stands for the frequency. The solid line represents a truncated power law, $\tau^{-\alpha}\exp(-\tau/\tau_0)$, with $\alpha = 1.3$ and $\tau_0 = 76$ min.

## 7.  Movie watching

The data were collected by a large American company for mail orders of DVD-rental (www.netflix.com). The users can rate movies online. In total, the data comprises of 17,770 movies, 447,139 users and about 96.7 millions of records. Tracking the records of a given user $i$, one can get $k_i$ - 1 interevent times, where $k_i$ is the number of movies user $i$ has already watched. The time resolution of the data is one day. Figure 12.8 reports the interevent time distribution based on the aggregated data of all users. The distribution follows a power law for more than two orders of magnitude, with an exponent approximately equal to -2.08.

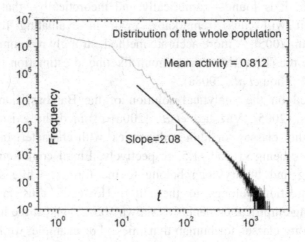

Fig. 12.8. Distribution of interevent time at the population level, indicating a power law [Zhou *et al.*, 2008a]. The solid line in the log-log plot has slope -2.08. The data exhibits weekly oscillations, reflecting a weekly periodicity of human behavior, which has also been observed in email communications [Holme, 2003].

Although the Poisson process is widely used to quantify the consequences of human actions, yet, an increasing number of empirical results indicate non-Poisson statistics of the timing of many human actions. Specifically, the interevent time $\tau$ or the response time $\tau_w$ obeys a heavy-tailed distribution. Besides what have been described above, more evidence could be found in the on-line games [Henderson & Nhatti, 2001], the Internet chat [Dewes *et al.*, 2003], FTP requests initiated by individual users [Paxson & Floyd, 1996], timing of printing jobs submitted by users [Harder & Paczuski, 2006], and so on. While the majority of these heavy tails can actually be well approximated by a power law, there also exist debates on the choice of the fitting function in the case of interevent time distribution in email communications [Stouffer *et al.*, 2005; Barabási *et al.*, 2005]. Another candidate, the log-normal distribution, has been suggested [Stouffer *et al.*, 2005] for describing the non-Poisson temporal statistics of human activities. After choosing a power law as the fitting function, the next problem is how to evaluate its exponent [Goldstein *et al.*, 2004]. In most academic reports, the exponent is directly obtained by using a linear fit in the log-log plot. However, it is found—numerically and theoretically—that this simple linear fit will cause remarkable error in evaluating the exponent [Newman, 2005]. A more accurate method, strongly recommended by us, is to use the (logarithmic) maximum likelihood estimation [Goldstein *et al.*, 2004; Zhou *et al.*, 2008a].

Based on the analytical solution of the Barabási queuing model [Barabási, 2005], Vázquez *et al.* [2006] claimed the existence of two universality classes for human dynamics, with characteristic power-law exponents being -1 and -1.5, respectively. Email communication, web browsing and library loan belong to the former, while surface mail communication belongs to the latter. However, thus far, there are increasing empirical evidence, as shown above, against the hypothesis of universality classes for human dynamics. For example, we [Zhou *et al.*, 2008a] sort the Netflix users by activity (i.e., the frequency of events of an individual, here meaning the frequency of movie ratings) in a descending order, and then divide this list into twenty groups, each of which has almost the same number of users. As shown in Fig. 12.9, we observe a monotonous relation between the power-law exponent and the mean activity in the group, which suggests that the activity of individuals is one of the key ingredients determining the distribution of interevent times. And the tunable exponents controlled by a single parameter indicate a far different scenario against the discrete universality classes suggested by Vázquez *et al.* [2006]. A similar relationship between activity and power-law exponent is also reported in the analysis of short-message communication [Hong *et al.*, 2008].

In addition, Vázquez *et al.* [2006] suggest that the waiting time distribution of tasks could in fact drive the interevent time distribution, and that the waiting time and the interevent time distributions should decay with the same scaling exponent. Our empirical study on correspondence pattern [Li *et al.*, 2007] supports this claim. However, the real situation should be more complicated, and a solid conclusion is not yet achieved.

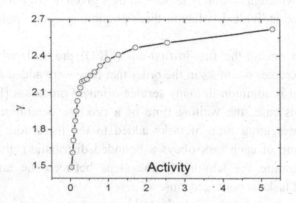

Fig. 12.9. Dependence of power-law exponent γ of interevent time distribution on mean activity of each group [Zhou *et al.*, 2008a]. Each data point corresponds to one group. All the exponents are obtained by using maximum likelihood estimation [Goldstein *et al.*, 2004] and can pass the Kolmogorov-Smirnov test with threshold quantile 0.9.

In short, abundant and in-depth empirical analyses are required before a completely clear picture about the temporal statistics of human-initiated systems can be drawn.

## 12.3 The Task-Driven Model

What is the underlying mechanism leading to such a human activity pattern? One potential starting point to arrive at an answer is to consider the queuing of tasks. A person needs to face many works in his/her daily life, such as sending email or surface mail, making telephone call, reading papers, writing articles, and so on. Generally speaking, in our daily life, we are doing these works one by one with some kind of order. In the modeling of human behaviors, we can abstract these activities on human life as tasks. Accordingly, Barabási [2005] proposed a model based on queuing theory.

In this model [Barabási, 2005], an individual is assigned a list with $L$ tasks. The length of list mimics the capacity of human memory for tasks waiting for execution. At each time step, the individual chooses a task from the list to execute. After being executed, it is removed from the list, and a new task is added. Each task is assigned a priority parameter $x_i$ ($i$ = 1, 2, $\cdots$, $L$), which is randomly generated by a given distribution function $\eta(x)$. Here the individual is facing three possible selection protocols for these tasks:

The first one is the first-in-first-out (FIFO) protocol, wherein the individual executes the tasks in the order that they were added to the list. This protocol is common in many service-oriented processes [Reynolds, 2003]. In this case, the waiting time of a task are determined by the cumulative executing time of tasks added to the list before it. If the executing time of each task obeys a bounded distribution, the waiting time, representing the length of time steps between the arrival and execution, of tasks is homogeneous.

The second one is to execute the tasks in a random order independent of their priority and arriving time. In this case, the waiting time distribution of tasks is exponential [Gross & Harris, 1985].

The last but most important one is the highest-priority-first (HPF) protocol. In this case, the tasks with highest priority are executed first, even though they are added later in the list. Hence, the tasks with lower priority could wait for a long time before being executed. Such protocol exists widely in human behaviors; for instance, we usually do the most important or the most urgent works first, and then the others.

The Barabási model [2005] focuses on the effect of the HPF protocol. At each time step, it assumes that the individual executes the task with the highest priority with probability $p$, and executes a randomly chosen task with probability $1 - p$. Obviously, if $p \to 0$, the model obeys the second protocol, and if $p \to 1$, it displays a pure HPF protocol.

Simulation results with $\eta(x)$ being a uniform distribution in [0, 1] are shown in Fig. 10.10. For $p \to 0$ (random chosen protocol), the waiting time distribution $P(\tau)$ decays exponentially; for $p \to 1$ (HPF protocol), it follows a power-law distribution with exponent -1, which agrees well with the empirical data of email communication. The results shown in Fig. 12.10a are generated with list length $L = 100$; however, the tail of the waiting time distribution $P(\tau)$ is independent of $L$, and the observed heavy-tailed property holds even for $L = 2$. Its exact analysis for the case $L = 2$ and different $p$ is given in [Vázquez, 2005; Vázquez *et al.*, 2006; Gabrielli & Caldarelli, 2007]. As shown in Fig. 12.11, it is not necessary to have a long priority list for individuals. If an individual can balance at least two tasks, the heavy-tailed property of the waiting time distribution will emerge. These results imply that the HPF protocol could be an important mechanism leading to the non-Poisson statistics of human dynamics.

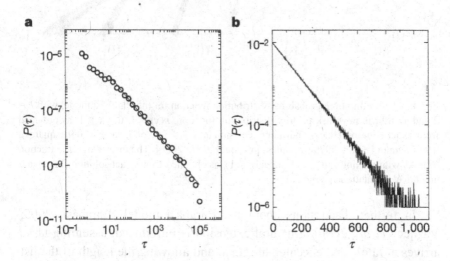

Fig. 12.10. Waiting time distribution predicted by the queuing model [Barabási, 2005]. Priorities were chosen from a uniform distribution $x_i \in [0, 1]$, and numerical simulation monitors a priority list of length $L = 100$ over $T = 10^6$ time steps. **a,** Log-log plot of the tail of probability $P(\tau)$ that a task spends $t$ time on the list obtained for $p = 0.99999$ (where p is the probability that the highest-priority task will be first executed) corresponding to the deterministic limit of the model. The straight line in the log–log plot has slope -1, in agreement with numerical results and analytical predictions. The data were log-binned, to reduce the uneven statistical fluctuations common in heavy-tailed distributions, a procedure that does not alter the slope of the tail. **b,** Linear-log plot of the $P(\tau)$ distribution for $p = 0.00001$, corresponding to the random choice limit of the model. The fact that the data follows a straight line on a linear-log plot indicates that $P(\tau)$ decays exponentially.

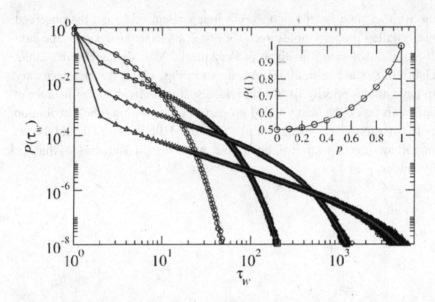

Fig. 12.11. Waiting time probability distribution function for the Barabási model with $L = 2$ and a uniform new-task priority distribution function, $\eta(x) = 1$, $0 \leq x \leq 1$, as obtained from exact solution (lines) and numerical simulations (symbols), for $p = 0.9$ (squares), 0.99 (diamonds) and 0.999 (triangles) [Vázquez et al., 2006]. The inset shows the fraction of tasks with waiting time $\tau = 1$, namely $P(1)$, as obtained from exact solution (lines) and numerical simulations (symbols).

In further discussions about the Barabási model [Barabási, 2005; Vázquez et al., 2006], a natural extension is introduced: assuming tasks arrives at rate $\lambda$ and executed at rate $\mu$, and allowing the length of the list of tasks to change in time. Let $\rho = \lambda/\mu$; obviously, here are three different cases need to be discussed [Vázquez et al., 2006]:

The first case is the *subcritical* regime, where $\rho < 1$, namely the arrival rate is smaller than the execution rate. In this case, the list will be often empty, and most tasks are executed soon after their arrival, thus the long term waiting time is limited. Simulations indicate that the waiting time distribution exhibits an exponential decay when $\rho \to 0$, and when $\rho \to 1$ it is close to a power-law distribution with exponent -3/2 and an exponential cutoff.

The second case is the *critical* regime, where $\rho = 1$. Here, the length of the list is a random walk in time. Different from the case with a fixed

$L$, the fluctuation in the list length will affect the waiting time distribution. Simulation results indicate that the waiting time distribution obeys a power law with exponent -3/2.

The third case is the *supercritical* regime, where $\rho > 1$, namely the arrival rate is larger than the execution rate. Thus the length of the list will grow linearly, and a $1 - 1/\rho$ fraction of tasks are never executed. Simulations indicate that the waiting time distribution in this case obeys a power law with exponent -3/2, too. A problem is how to understand such a growing list. Let us think over the case of replying regular mails. When we receive a mail, we put it on desk and piled it up with early mails, and we usually choose a mail from the pile to reply. If the received mails are too many and we do not have enough time to reply all of them, the mails in the pile will become more and more. We do not need to remember the list, because all the mails are put on there. Therefore, the list of mails waiting for reply is unlimited. Simulation results for this case are in agreement with the empirical data of surface mail replies of Darwin, Einstein and Freud. All these three great scientists have many mails never being replied. Numerical results are shown in Fig. 12.12 [Vázquez *et al.*, 2006], which is in accordance with the above analysis.

The above model only considers the behavior of an isolated individual. Actually, every person is living in a surrounding society with countless interactions to others; such interactions may affect our activities, such as email communication, phone calling and all collaborated works. In a recent model on human dynamics, based on the queuing theory, the simplest case taking into account the interactions between only two individuals is considered [Oliveira & Vázquez, 2007]. This model only considers two individuals: A and B. Each individual has two kinds of tasks, an interacting task (I) that must be executed in common, and a aggregated non-interacting task (O) that can be executed by the individual himself/herself. Each task is assigned a random priority $x_{ij}$ ($i = $ I, O; $j = $ A, B) extracted from a probability density function $\eta(x)$. At each time step, both agents select a task with highest priority in their list (with length $L_A$ for A and $L_B$ for B). If both agents select task I, then it is executed; otherwise each agent executes a task of type O.

As shown in Fig. 12.13, numerical simulations of this model indicate that the interevent time distribution of the interacting task I is close to a power law, having a wide range of exponents. This result extends the range of the Barabási model, and highlights a potential way to understand the pattern of the non-Poisson statistics in the interacting activities of human.

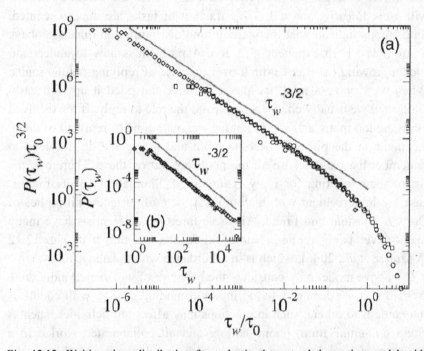

Fig. 12.12. Waiting time distribution for tasks in the extended queuing model with continuous priorities [Vázquez *et al.*, 2006]. Numerical simulations were performed as follows. At each step, the model generates an arrival $\tau_a$ and service time $\tau_s$ from an exponential distribution with rates $\lambda$ and $\mu$, respectively. If $\tau_a < \tau_s$ or there are no tasks in the queue, a new task is then added to the queue, with a priority $x \in [0, 1]$ from uniform distribution, and update the time $t \to t + \tau_a$. Otherwise, the model removes from the queue the task with the largest priority and update the time $t \to t + \tau_s$. The waiting time distribution is plotted for three $\rho$ (= $\lambda/\mu$) values: $\rho = 0.9$ (circles), 0.99 (squares) and 0.999 (diamonds). The data has been rescaled to emphasize the scaling behavior $P(\tau_w) = \tau_w^{-3/2} f(\tau_w / \tau_0)$, where $\tau_0 \sim (1 - \rho^{1/2})^{-2}$. The inset shows the distribution of waiting times for $\rho = 1.1$, after collecting up to $10^4$ (plus) and $10^5$ (diamonds) *executed* tasks, showing that the distribution of waiting times has a power law tail even for $\rho > 1$ (supercritical regime). Note that in this regime a high fraction of tasks are never executed, staying forever on the priority list whose length increases linearly with time, a fact that is manifested by a shift to the right of the cutoff of the waiting time distribution.

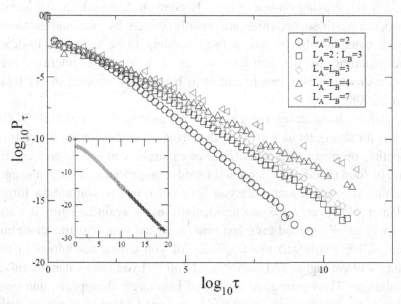

Fig. 12.13. Probability distribution of interevent time $\tau$ of the interacting task I, as obtained from direct numerical simulations of the model [Oliveira & Vázquez, 2007]. Each data set was obtained after $10^{11}$ model time steps, corresponding to total number of I plus O task executions. Note that as $L_A$ and/or $L_B$ increases it becomes computationally harder to have a good estimate of $P$ because the execution of the I task becomes less frequent.

## 12.4 The Interest-Driven Model and Beyond

The motivations of our behaviors are extremely complex [Kentsis, 2006]. Therefore, setting up the rules in modeling by simplifying or coarse-graining the real world is the main way (even the only possible way) in the study of human dynamics [Oliveira & Barabási, 2006]. Although the queuing models do obtain a great success in explaining the heavy tails in human dynamics, they have their own limitations. Actually, the core and fundamental assumption of the queuing models is that the behaviors of humans are treated as executing tasks; however, this assumption could not fit all human behaviors. Some real-world human activities could not be explained by a task-based mechanism, but they could still exhibit similar statistical law (heavy-tailed interevnt time distribution), such as browsing webs [Dezsö *et al.*, 2006], watching on-line movies [Zhou *et al.*, 2008a], playing on-line games [Henderson & Nhatti, 2001], and so on. Clearly, these activities are mainly driven by personal interests, which could not be treated as tasks needing to be executed. In-depth understanding of the non-Poisson statistics in these interest-driven systems requires new insight and ideas beyond the queuing theory [Han *et al.*, 2008].

Before introducing the rules of an interest-driven model (HZW model for short), let us think over the changing process of our interests or appetites on many daily activities. For example, you had eaten a certain kind of food (hamburger, say) and found it tasting good a long time ago. Because for such a long time you have not tasted it, you almost forgot about it. However, after you accidentally eat it again and find it tasty, you will recall its good taste last time, and then your appetite about this food will get stronger; the frequency of you eating hamburger in the future will get higher and higher, until you feel you have eaten too much hamburger. Then your good feeling of hamburger disappears, and your appetite for it also weakens after a long period of time. A similar daily experience can be found in Web browsing. If a person has not browsed the Web in a long time, an accidental browsing event may give him a good feeling and wake up his old interest on Web browsing. Next, during the activity the good feeling is durative, and the frequency of Web browsing in future may increase. Then, if the frequency is too high, he may worry about the habit and reduce those browsing activities. We can also find similar changing processes of interest or appetite in many other daily activities, such as game playing, movie watching, and so on. In short, we usually adjust the frequency of our daily activities according to our interest: greater interest will lead to higher frequency, vice versa. In other words, our interests are history dependent or adaptively changing.

To mimic these daily experiences, the following simple assumptions in the modeling of interest-driven systems are extracted:

1. Each activity will change the current interest for a give interest-driven behavior, while the frequency of activities depends on the interest

2. We assume the interevent time $\tau$ has two thresholds: when $\tau$ is too small (that is, events happen too frequently), interest will be

depressed and thus the interevent time will increase; while if the time gap is too long, we will impose an occurrence to mimic a casual action.

According to the assumptions above, the rules of the HZW model are listed as follows:

1. Time is discrete and labeled by $t = 0, 1, 2,...$; the occurrence probability of an event at time step $t$ is denoted by $r(t)$. The time interval between two consecutive events is call the interevent time and denoted by $\tau$.
2. If the $(i+1)^{th}$ event occurred at time step $t$, the value of $r$ is updated as $r(t + 1) = a(t)r(t)$, where $a(t) = a_0$ if $\tau_i \leq T_1$, $a(t) = a_0^{-1}$ if $\tau_i \geq T_2$, and $a(t) = a(t - 1)$ if $T_1 < \tau_i < T_2$.

If no event occurs at time step $t$, we set $a(t) = a(t - 1)$; namely $a(t)$ remains unchanged. Here $T_1$ and $T_2$ are two thresholds satisfied by $T_1 << T_2$; $\tau_i$ denotes the time interval between the $(i + 1)^{th}$ and the $i^{th}$ events; $a_0$ is a parameter controlling the changing rate of occurrence probability ($0 < a_0 < 1$). If no event happens, the value of $r$ will not change. Clearly, simultaneously enlarging (with the same multiplication factor) $T_1$, $T_2$ and the minimal perceptible time will not change the statistics of this system. Therefore, without lost of generality, the lower boundary $T_1$ is set to 1.

In the simulations, we set the initial interest, $r_0 = r(0)$, equal to 1, which is also the maximum value of $r(t)$ in the whole process. As shown in Fig. 12.14, the succession of events predicted by the HZW model exhibits very long inactive periods separated by bursts of rapidly occurring events, and the corresponding $r(t)$ shows clearly seasonal changing property. Fig. 12.15 presents simulation results with tunable $T_2$ and $a_0$. Given $a_0 = 0.5$, if $T_2 >> T_1$, the interevent time distribution generated by the present model displays a power law with exponent -1; while if $T_2$ is not sufficiently large, the distribution $P(\tau)$ will depart from the power law, exhibiting a cutoff in the tail. Correspondingly, given sufficiently large $T_2$, the effect of $a_0$ is very slight, which can thus be ignored. The power-law exponent -1 can also be analytically obtained under the circumstance of $T_2 \gg T_1$; detailed mathematical derivation can be found in [Han *et al.*, 2008].

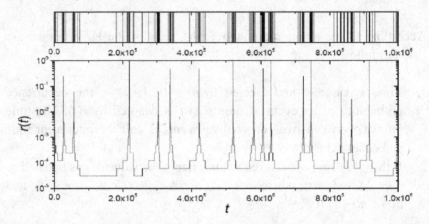

Fig. 12.14. *Upper panel*: Succession of events predicted by the HZW model. Total number of events shown here is 375 during $10^6$ time steps. *Lower panel*: The corresponding changes of $r(t)$. Data points are obtained with the parameters $a_0 = 0.5$ and $T_2 = 10^4$. [Han *et al.*, 2008]

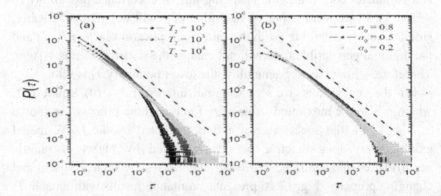

Fig. 12.15. Interevent time distributions in log-log plots [Han *et al.*, 2008]. (a) Given $a_0 = 0.5$, $P(\tau)$ for different $T_2$, where the black, dark gray and bright gray curves denote the cases of $T_2 = 10^2$, $10^3$ and $10^4$, respectively. (b) Given $T_2 = 10^4$, $P(\tau)$ for different $a_0$, where the black, dark gray and bright gray curves denote the cases of $a_0 = 0.8$, 0.5 and 0.2, respectively. The black dash lines in both (a) and (b) have slope -1. All data points are obtained by averaging over 100 independent runs, and each includes $10^4$ events.

Differing from the queuing models discussed in Section 12.3, this model is driven by the personal interest. In this model, the frequency of events is determined by the interest, while the interest is in turn affected by the occurrence of events. This intertwining mechanism, similar to that in an active walk[5] [Lam, 2005; 2006], is *a generic origin of complexity* in many real-life systems. The rules of the model are extracted from the daily experiences of people, and the simulation results agree with many empirical observations, such as the activities of Web browsing [Dezsö *et al.*, 2006]. This model indicates a much simple activity pattern of human behaviors; that is, people could adaptively adjust their interest on a specific behavior (for example, TV watching, Web browsing, on-line game playing, etc.), which leads to a quasi-periodic change of interest, and this quasi-periodic property eventually gives rise to the departure from Poisson statistics. This simple activity pattern could be universal in many human behaviors.

There are also many other models which are neither based on queuing theory nor the interest-driven mechanism. One important model, proposed by Vázquez [2007], takes into account the memory of past activities by assuming that a person reacts by accelerating or reducing their activity rate based on the perception of their past activity rate. Let $\lambda(t)dt$ be the probability the individual performs the activity between time $t$ and $t + dt$. Based on this assumption, the equation for $\lambda(t)$ can be written as follows:

$$\lambda(t) = a\frac{1}{t}\int_0^t dt\lambda(t) \tag{12.3}$$

where $a > 0$ is the only parameter in this model. When $a = 1$, $\lambda(t) = \lambda(0)$ and the process is stationary. On the other hand, when $a \neq 1$ the process is non-stationary with acceleration ($a > 1$) or reduction ($a < 1$).

Note that Eq. (12.3) includes a latent assumption on the starting time ($t = 0$). As indicated in [Vázquez, 2007], it is a reflection of our bounded memory, meaning that we do not remember or do not consider what took

---

[5] In an active walk, a particle (the walker) is coupled to a deformable landscape, influencing each other as the particle moves or the landscape changes.

place before that time. For instance, we usually check for new emails every day after arriving at work no matter what we did the day before.

From Eq. (12.3), one can obtain the function of interevent time distribution via mathematical derivations; details are given in [Vázquez, 2007]. The general conclusion of the model is as follows: When $a = 1$, the resulting interevent time distribution is an exponential function. Let $\tau_0 = (a\lambda_0)^{-1}$, where $\lambda_0$ is the mean number of events in the considered time period $T$. When $a > 1$ (acceleration) and $\tau_0 \ll \tau < T$, the interevent time distribution generated by the model is close to a power law with exponent $-2 - (a - 1)^{-1}$. On the other hand, when $0 < a < 1/2$ (reduction) and $\tau \ll \tau_0$, its interevent time distribution is close to a power law with exponent $-1 + a/(1 - a)$.

Comparing with the cumulative number of regular mails sent out by Darwin and Einstein as a function of time and the interevent time distribution of these regular mails, results generated by this simple model are in agreement with the empirical data, as shown in [Vázquez, 2007].

## 12.5 Discussion and Conclusion

The study of human dynamics, still in its infancy, is currently attracting a lot of attention due to their theoretical and practical significances. Extensive empirical evidence demonstrates non-Poisson statistics in temporal human activities, in opposite to what is expected from the traditional hypothesis [Gross & Harris, 1985] which assumes uniform and stationary timing of human actions. Actually, the majority of real applications of queuing theory in the past are based on the assumption of a Poisson process of event occurrence [Newell, 1982]. From now on, we may have to complement queuing theory by taking into account the possibility of a power-law interevent time distribution. Since the second moment of a power-law distribution with exponent in the interval $[-3, \infty)$ is *not* convergent, many previous conclusions in queuing theory are invalid when a heterogeneous interevent time distribution is considered. Mathematically speaking, the new findings on non-Poisson temporal statistics give rise to an important open question in queuing theory. Theoretical interests aside, these new findings have significant potential in practical applications [Barabási, 2005]. For example, in-depth understanding of human activity pattern is indispensable in modeling social structures [Zhu et al., 2008] and financial behaviors [Caldarelli et al., 1997], and is also crucial for better resource allocation and pricing plans for telephone companies [Barabási, 2005].

Thus far, many models aiming at the explanation of the origin of heavy-tailed human activity pattern have been proposed. The majority of previous models before our HZW model are based on queuing theory. Yet, not all human-initiated systems are driven by some tasks. Besides the task-driven mechanism underlying the queuing theory, some other possible origins of human behavior—such as interest [Han et al., 2008] and memory [Vázquez, 2007]—are also highlighted recently. As stated by Kentsis [2006], there are countless number of ingredients affecting human behaviors, and for most of them, we do not know their impacts. We believe, in the near future, more theoretical models will be proposed to reveal the effects of task deadline, task optimization protocol, human seasonality, social interactions, and so on.

Another important issue is how the non-Poisson temporal statistics affect the relative dynamical processes taking place in human-initiated systems. For example, epidemic spreading of diseases, such as AIDS, influenza and SARS, is driven by social contacts between infected and susceptible persons. At the macroscopic level, the effects of social structures (i.e., epidemic contact networks) have been extensively investigated (see review article [Zhou et al., 2006a] and references therein). However, these works lack a serious consideration of the microscopic factor, namely the temporal statistics of epidemic contacts. Actually, prior works either assume the contact frequency of a given person being proportional to his/her social connectivity [Pastor-Satorras & Vespignani, 2003] or assume the same contact frequency for all infected persons [Zhou et al., 2006b]. Recently, based on the statistical reports of email worms, Vázquez et al. [2007] studied the impact of non-Poisson activity patterns on epidemic spreading processes; this work provides a starting point in understanding the role of individual activity pattern in aggregated dynamics.

In addition, heterogeneous traffic-load distribution as well as long-range correlation embedded in the traffic-load time series have been observed in many human-initiated systems, such as the Internet traffic

[Park & Willinger, 2000; Zhou *et al.*, 2006] and air transportation [Guimerà *et al.*, 2005; Liu & Zhou, 2007]. The observed heavy-tailed timing of email communication [Barabási *et al.*, 2005] and Web browsing [Dezsö *et al.*, 2006] as well as long range human travel [Brockmann *et al.*, 2006] may contribute to non-trivial phenomena. Furthermore, some social dynamics may also be highly affected by human activity patterns [Castellano *et al.*, 2007].

All previous studies on human dynamics focus on the distributions of interevent time and response time. However, its methodology—extracting statistical laws from historical records of human activities—is not limited to this issue. For instance, we can also use some of those data sets to quantify the herd behavior of an individual; that is, a person follows the opinions of the majority of people in his/her social surrounding in an irrational way. Although the herd behavior has found its significant impact on financial markets [Bikhchandani & Sharma, 2000], it is very hard to be quantified outside the laboratorial surrounding [Asch, 1955]. In many Web-based recommender systems, such as the on-line movie-sharing system [Zhou *et al.*, 2008b], the user's records contain not only the time he/she watch the movies, but also his/her opinion (i.e., ratings) on those movies. Similar records can also be found in the systems of Web-based trading, book-sharing, music-sharing, and so on. Since before the vote is cast by the user, he/she can see previous ratings assigned by others to this item, the herd behavior may occur. Quantitatively uncovering the latent bias of human opinion is crucial for better design of recommender systems [Zhang *et al.*, 2007; Zhou *et al.*, 2007; 2008a; 2008b; 2008c; Ren, 2008].

## References

Asch, S. E. [1955] "Opinion and social pressure," *Sci. Am.* **193**(5), 31-35.
Barabási, A.-L. [2005] "The origin of burst and heavy tails in human dynamics," *Nature* **435**, 207-211.
Barabási, A.-L., Goh, K.-I. & Vázquez, A. [2005] "Reply to Comment on 'The origin of bursts and heavy tails in human dynamics'," arXiv: physics/0511186.
Barabási, A.-L. [2007] "The architecture of complexity: From network structure to human dynamics," *IEEE Control Systems Magazine* **27**, 33-42.

Bikhchandani, S. & Sharma, S. [2000] "Herd behavior in financial markets: A review," *IMF Staff Papers* **47**, 279-310.

Brockmann, D., Hufnagel, L. & Geisel, T. [2006], "The scaling laws of human travel," *Nature* **439**, 462-465.

Caldarelli, G., Marsili, M. & Zhang, Y. C. [1997] "A prototype model of stock exchange," *Euro. Phys. Lett.* **40**, 479-484.

Castellano, C., Fortunato, S. & Loreto, V. [2007] "Statistical physics of social dynamics," arXiv: 0710.3256.

Dewes, C., Wichmann, A. & Feldman, A. [2003] "An analysis of Internet chat systems," *Proc. 2003 ACM SIGCOMM Conf. on Internet Measurement* (ACM, New York).

Dezsö, Z., Almaas, E., Lukács, A., Rácz, B., Szakadát, I. & Barabási, A.-L. [2006] "Dynamics of information access on the Web," *Phys. Rev. E* **73**, 066132.

Eckmann, J. P., Moses, E. & Sergi, D. [2004] "Entropy of dialogues creates coherent structure in email traffic," *Proc. Natl. Acad. Sci. U.S.A.* **101**, 14333-14337.

Gabrielli, A. & Caldarelli, G. [2007] "Invasion percolation and critical transient in the Barabási model of human dynamics," *Phys. Rev. Lett.* **98**, 208701.

Goldstein, M. L., Morris, S. A. & Yen, G. G. [2004] "Problems with fitting to the power-law distribution," *Eur. Phys. J. B* **41**, 255-258.

Gross, D. & Harris, C. M. [1985] *Fundamentals of Queuing Theory* (Wiley, New York).

Guimerà, R., Mossa, S., Turtschi, A. & Amaral, L. A. N. [2005] "The worldwide air transportation network: Anomalous centrality, community structure, and cities' global roles," *Proc. Natl. Acad. Sci. U.S.A.* **102**, 7794 – 7799.

Haight, F. A. [1967] *Handbook of the Poisson Distribution* (Wiley, New York).

Han, X.-P., Zhou, T. & Wang, B.-H. [2008] "Human dynamics with adaptive interest," *New J. Phys.* **10**, 073010.

Harder, U. & Paczuski, M. [2006] "Correlated dynamics in human printing behavior," *Physica A* **361**, 329 – 336.

Henderson, T. & Nhatti, S. [2001] "Modelling user behavior in networked games," *Proc. 9th ACM Int. Conf. on Multimetia 212–220* (ACM, New York).

Hidalgo, R. C. A. [2006] "Conditions for the emergence of scaling in the inter-event time of uncorrelated and seasonal systems," *Physica A* **369**, 877 – 883.

Holme, P. [2003] "Network dynamics of ongoing social relationships," *Euro. Phys. Lett.* **4**, 427-433.

Hong, W., Han, X.-P., Zhou, T. & Wang, B.-H. [2008] "Scaling behaviors in short-message communications," arXiv: 0802.2577.

Kentsis, A. [2006] "Mechanisms and models of human dynamics: Comment on 'Darwin and Einstein correspondence patterns'," *Nature* **441**, E5.

Lam, L. [2005] "Active walks: The first twelve years (Part I)," *Int. J. Bifurcation & Chaos* **15**, 2317-2348.

Lam, L. [2006] "Active walks: The first twelve years (Part II)," *Int. J. Bifurcation & Chaos* **16**, 239-268.

Li, N. N., Zhang, N. & Zhou, T [2007] "Empirical analysis on human correspondence pattern" (unpublished).

Liu, H.-K. & Zhou, T. [2007] "Topological properties of Chinese city airline networks," *Dynamics of Continuous, Discrete and Impulsive Systems B* **14**(S7), 135-138.

Newell, G. F. [1982] *Applications of Queuing Theory* (Chapman and Hall, New York).

Newman, M. E. J. [2005] "Power laws, Pareto distributions and Zipf's law," *Contemp. Phys.* **46**, 323-351.

Oliveira, J. G. & Barabási, A.-L. [2005] "Darwin and Einstein correspondence patterns," *Nature* **437**, 1251.

Oliveira, J. G. & Barabási, A.-L. [2006] "Reply to Comment on 'Darwin and Einstein correspondence patterns'," *Nature* **441**, E5-E6.

Oliveira, J. G. & Vázquez, A. [2007] "Impact of interactions on human dynamics," arXiv: 0710.4916.

Park, K & Willinger, W. [2000] "Self-Similar network traffic: An overview," in *Self-Similar Network Traffic and Performance Evaluation*, eds. Park, K. & Willinger, W. (Wiley, New York).

Pastor-Satorras, R. & Vespignani, A. [2003] "Epidemics and immunization in scale-free networks," in *Handbook of Graph and Networks*, eds. Bornholdt, S. & Schuster, H. G. (Wiley-VCH, Berlin).

Paxson, V. & Floyd, S. [1996] "Wide-area traffic: The failure of Poisson modeling," *IEEE/ACM Trans. Netw.* **3**, 226-244.

Ren, J., Zhou, T. & Zhang, Y. C. [2008] "Information filtering via self-consistent refinement," *Euro. Phys. Lett.* **82**, 58007.

Reynolds, P. [2003] *Call Center Staffing* (Call Center School, Lebanon, TN).

Scalas, E., Kaizoji, T., Kirchler, M., Huber, J. & Tedeschi, A. [2006] "Waiting times between orders and trades in double-auction markets," *Physica A* **366**, 463-471.

Stouffer, D. B., Malmgren, R. D. & Amaral, L. A. N. [2005] "Comment on 'The origin of bursts and heavy tails in human dynamics'," arXiv: physics/0510216.

Vázquez, A. [2005] "Exact Results for the Barabási Model of Human Dynamics," *Phys. Rev. Lett.* **95**, 248701.

Vázquez, A., Oliveira, J. G., Dezsö, Z., Goh, K.-I., Kondor, I. & Barabási, A.-L. [2006] "Modeling burst and heavy tails in human dynamics," *Phys. Rev. E* **73**, 036127.

Vázquez, A. [2007] "Impact of memory on human dynamics," *Physica A* **373**, 747-752.

Vázquez, A., Rácz, B., Lukács, A. & Barabási, A.-L. [2007] "Impact of non-

Poissonian activity patterns on spreading processes," *Phys. Rev. Lett.* **98**, 158702.

Watson, J. B. [1913] "Psychology as the behaviorist views it," *Psychological Review* **20**, 158-177.

Zhang, Y. C., Medo, M., Ren, J., Zhou, T., Li, T. & Yang, F. [2007] "Recommendation model based on opinion diffusion," *Euro. Phys. Lett.* **80**, 68003.

Zhao, G. S., Zhang, N. & Zhou, T. [2008] "Web browsing patterns of grouped users" (unpublished).

Zhou, P.-L., Cai, S.-M., Zhou, T. & Fu Z.-Q. [2006] "Scaling behaviors of traffic in computer communication networks," *2006 International Conference on Communications, Circuits and Systems Proceedings* (IEEE, New York) pp.2687-2691.

Zhou, T., Fu, Z.-Q. & Wang, B.-H. [2006a] "Epidemic dynamics in complex networks," *Prog. Nalt. Sci.* **16**, 452-457.

Zhou, T., Liu, J.-G., Bai, W.-J., Chen, G.-R. & Wang, B.-H. [2006b] "Behaviors of susceptible-infected epidemics in scale-free networks with identical infectivity," *Phys. Rev. E* **74**, 056109.

Zhou, T., Ren, J., Medo, M. & Zhang, Y. C. [2007] "Bipartite network projection and personal recommendation," *Phys. Rev. E* **76**, 046115.

Zhou, T., Jiang, L.-L., Su, R.-Q. & Zhang, Y. C. [2008a] "Effect of initial configuration on network-based recommendation," *Euro. Phys. Lett.* **81**, 58004.

Zhou, T., Kiet, H. A.-T., Kim, B. J., Wang, B.-H. & Holme, P. [2008b] "Role of activity in human dynamics," *Euro. Phys. Lett.* **82**, 28002.

Zhou, T., Su, R.-Q., Liu, R.-R., Jiang, L.-L., Wang, B.-H. & Zhang, Y. C. [2008c] "Ultra accurate personal recommendation via eliminating redundant correlations," arXiv: 0805.4127.

Zhu, C.-P., Zhou, T., Yang H.-J., Xiong, S.-J., Gu, Z.-M., Shi, D.-N., He, D.-R. & Wang, B.-H. [2008] "The process of co-evolutionary competitive exclusion: Speciation, multi-factuality and power laws in correlation," *New J. Phys.* **10**, 023006.

# 13

## Human History: A Science Matter

*Lui Lam*

Human history is the most important discipline of study. The complex system under study in history is a many-body system consisting of *Homo sapiens*—a (biological) material system. Consequently, history is a legitimate branch of science, since science is the study of Nature which includes *all* material systems. A historical process, expressed in the physics language, is the time development of a subset of or the whole system of *Homo sapiens* that happened during a time period of interest in the past. History is therefore the study of the past dynamics of this system. Historical processes are stochastic, resulting from a combination of contingency and necessity. Here, the nature of history is discussed from the perspective of complex systems. Human history is presented as an example of Science Matters. Examples of various scientific techniques in analyzing history are given. In particular, two unsuspected *quantitative* laws in Chinese history are shown. Applications of active walks to history are summarized. The "differences" between history and the natural sciences erroneously expressed in some history textbooks are clarified. The future of history, as a discipline in the universities, is discussed; recommendations are provided.

### 13.1  What is History?

Human history is the most important discipline of study [Lam, 2002]. Yet, human history, or history in general, as a science is rarely discussed [Lam, 2002; Krakauer, 2007].

Science is the study of Nature and to understand it in a unified way. Nature, of course, includes all material systems. The system investigated in history is a (biological) material system consisting of *Homo sapiens*. Consequently, history is a legitimate branch of science, like physics, biology and paleontology, and so on. In other words, history is not a subject that is beyond the domain of science. History can be studied scientifically [Lam, 2002].

By definition, history is about past events and is irreproducible. In this regard, it is like the other "historical" sciences such as cosmology, astronomy, paleontology and archeology. The way historical sciences advance is by linking them to systems presently exist, which are amenable to tests. For example, in astronomy, the color spectra of light emitted in the past from the stars and received on earth can be compared with those observed in the laboratory; the elements existing in stars is then identified. Similarly, the psychology, thoughts and behaviors of historical players can be inferred from those of living human beings, which can be learned by observations, experimentations and neurophysiologic probes [Feder, 2005].

The system under study in history is a many-body system. In this system, each "body" is a human being, called a "particle" here; these particles have internal states (due to thinking, memory, mood and so on) which sometimes can be ignored. Each constituent particle is a (non-quantum mechanical) classical object and is distinguishable; that is, each particle in the system can be identified individually. This many-body system is a heterogeneous system, due to the different sizes, ages, races...of the particles.

A historical process, expressed in the physics language, is the time development of a subset of or the whole system of *Homo sapiens* that happened during a time period of interest in the past. History is therefore the study of the past dynamics of this system. *Historical processes are stochastic, resulting from a combination of contingency and necessity.*[1] Here, necessity is an assumption, which could only be confirmed by results showing that it really exist; contingence is due to the many other factors not included in the system under study—as usually is the case in many complicated situations—and could be represented as noise in the study of stochastic systems.

---

[1] "Stochastic" is a technical word in physics, meaning that probability appears somewhere in the process; a random process is a special case [Paul & Baschnagel, 1999].

In modeling, contingency shows up as probability and necessity is represented by rules in the model. The situation is like that in a chess or soccer game. There are a few basic rules that the players have to obey, but because of contingency, the detail play-by-play of each game is different. In principle, someone with sufficient skills and patience can guess the rules governing historical processes, like those in a chess or soccer game.

In some cases, these two ingredients of contingency and necessity, through self-organization, may combine to give rise to discernable historical trends or laws. In other cases, either no laws exist at all or the laws are not recognized by whoever studying them. Whether there actually exist historical laws cannot be settled by speculations or debates, no matter how good these speculations or debates are. *A historical law exists only when it is found and confirmed* (as indeed is the case as shown in Sections 13.2.1 and 13.2.4). Furthermore, any historical law—like that in physics—has its own range of validity, which may cover only a limited domain of space and time. Yet most people, including many historians, do not believe that any historical law could exist [Gardiner, 1959]. They are wrong.

## 13.2 Methods to Study History

An important step towards the scientific study of any subject is to pick the right tool to tackle it. Historical processes are stochastic. The kind of physics suitable for handling many-body systems ingrained with contingency is statistical physics. Furthermore, the historical system is an *open* system with constant exchange of energy and materials with the environment and is never in equilibrium. Thus, for history, an appropriate tool is the stochastic methods developed in the statistical physics of nonequilibrium systems [Lam, 1998; Paul & Baschnagel, 1999; Sornette, 2000].

However, there are other tools, too. In fact, there are at least four different approaches applicable in understanding history. Examples are given below, with each reflecting either the *empirical, phenomenological* or *realistic* level commonly found in the scientific development of any discipline [Lam, 2002].[2]

## 13.2.1 *Statistical Analysis*

Statistical analyses of data are at the empirical level, without knowing the mechanism of the processes involved. Two examples are given here.

1.  Power law in the distribution of war intensities

Figure 13.1a shows a historical law of statistical nature; historical laws do exist. The statistical distribution of war intensities obeys a power law[3] (Fig. 13.1a), first discovered by Richardson [1941] and confirmed by Levy [1983] using a different data set covering 119 wars from 1495-1975. In this new study, war intensity is defined by the ratio of battle deaths to the population of Europe at the time of the war. (Europe is used because for the earlier wars, estimates of the world population are unreliable.) More recently, this conclusion is interpreted by Roberts and Turcotte [1998] in terms of a forest-fire model. Similar power law is found in the distribution of earthquake intensities, called Gutenberg-Richter law (Fig. 13.1b), in the ranking of city populations, and in many other systems [Zipf, 1949]. The fact that human events like wars obey the same statistical law as inanimate systems indicates that the human system does belong to a large class of dynamical systems in Nature, beyond the control of human intentions and actions, individually or collectively.

2.  Power law in the distribution of Chinese regime lifetimes

Another example is provided in the case of Chinese history. China has a long, unbroken history, which is probably the best documented [Huang, 1997]. The dynasties from Qin to Qing ranges from 221 BC to 1912, with 31 dynasties and 231 regimes spanning a total of 2,133 years [Morby, 2002]. (A regime is the reign of one emperor; a dynasty may

---

[2] For history, there is also the *artificial* level—artificial history [Lam, 2002].
[3] In a power law, two variables $x$ and $y$ relate to each other through $y = Ax^{\alpha}$, where $A$ and $\alpha$ are constants. Equivalently, the plot of $\log x$ vs. $\log y$ shows up as a straight line.

consist of several regimes.) Some of these dynasties overlap with each other in time.

Let $\tau_R$ be the regime lifetime, an integer measured in years. The histogram of $\tau_R$ is found to obey a power law (Fig. 13.2), with an exponent equal to -1.3 ± 0.5 [Lam, 2006a; 2006b]. This result implies that the dynamics governing regime changes is not completely up to the emperors, statistically speaking, but share some common traits with other complex systems such as those displayed in Fig. 13.1. To the best of our knowledge, this is *the first quantitative law concerning Chinese history.*

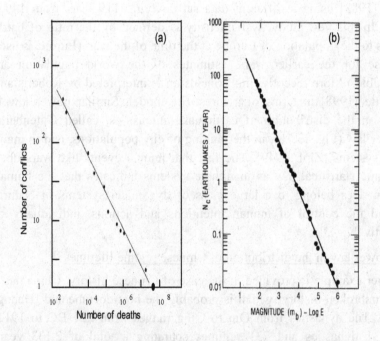

Fig. 13.1. (a) Statistical distribution of war intensities. Eighty-two wars from 1820 to 1929 are included; the dot on the horizontal axis comes from World War I. (b) Distribution of earthquake sizes in the New Madrid zone in the United States from 1974 to 1983 [Johnston & Nava, 1985]. The points show the number of earthquakes with magnitude larger than a given magnitude *m*. The graphs in (a) and (b) are log-log plots; a straight line indicates a power law.

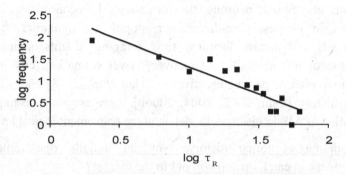

Fig. 13.2. Log-log plot of the histogram of Chinese regime lifetime $\tau_R$, covering the years from 221 BC to 1912. (Bin width of the histogram is 4 years.)

## 13.2.2 *Computer Modeling*

A common metaphor for history is that it is like a river flowing; people talk about the "river of history." This metaphor is not so off mark if the water flowing in the river is able to reshape the landscape as it flows, and the river is allowed to branch from time to time under certain conditions. Active walk (AW) [Lam, 2005a; 2006b] is a natural in matching such a metaphor. It is then no surprise that a whole class of probabilistic AW models are found to be relevant in studying history [Lam, 2002].

Active walk is a paradigm and method introduced by Lam in 1992 to handle self-organization and pattern formation in simple and complex systems. In AW, a particle (the walker) changes a deformable potential—the landscape—as it walks; its next step is influenced by the changed landscape. For example, ants are living active walkers. When an ant moves, it releases chemicals of a certain type and hence changes the spatial distribution of the chemical concentration. Its next step is moving towards positions of higher chemical concentration. In this case, the chemical distribution is the deformable landscape. Active walk has been applied successfully to a number of complex systems in "natural"[4] and

---

[4] In this chapter, "natural science" with quotation marks is defined as the science of mostly simple systems [Lam, 2008a].

social sciences. Examples include pattern formation in physical, chemical and biological systems such , as surface-reaction induced filaments and retinal neurons, the formation of fractal surfaces, ionic transport in glasses, granular matter, population dynamics, bacteria movements and pattern forming, food foraging of ants, spontaneous formation of human trails, oil recovery, river formation, city growth, economic systems, parameter networks [Han *et al.*, 2008] and, recently, human history [Lam, 2002; 2004; 2006b]. Here are some examples of application of AW in history—modeling at the phenomenological level.

1.  Modeling economic history: why an initially disadvantageous product can catch up and win out in the market?

A Florence cathedral clock, built in 1443, has hands that move *counterclockwise* around its dial [Arthur, 1990]. Consequently, two types of clocks, with hands moving in different directions, could be in the market in those early years. However, since sundials, the timepiece before the invention of clocks, in the north hemisphere have the pointer's shadow moving clockwise, people in Europe are more comfortable with clocks having hands moving clockwise, too. Those "counterclockwise" clocks, like the Florence cathedral clock, are thus "inferior" products and they are soon run out of the market. That is why we are now left with only one type of clocks, those having hands running clockwise. This is the case of an inferior product losing out, which is not at all surprising. What is surprising is the case of an inferior product winning out. The "QWERTY" keyboard, the type we are using today, is such an example [David, 1986]. Invented in 1867, this keyboard is designed to slow down our typing so the mechanical parts will not be jammed that easily. Other superior designs, such as the Maltron keyboard—with 91% of the letters used frequently in English on the "home row" compared to 51% for the QWERTY design—coexist in the market but all lose out. Why?

The two-site AW model [Lam, 2005a] is able to explain this and other real cases[5] in economic history that an inferior product can actually win out in the market [Lam, 2002].

---

[5] Other examples are the competition between Apple computers and the PCs, as well as Beta and VHS videotapes. In each case here, the second product is the inferior product.

2. Modeling evolutionary history: rewinding life's "tape," or how important is contingency in survival?

In 1989, Stephen Jay Gould (1941-2002) [1989] publishes the book *Wonderful Life*. From the fossil record found outside of Vancouver, Canada, it seems that some "advanced" organisms (with many legs, say) that should survive are wiped out suddenly. From this one data point, Gould concludes that contingency is extremely important, that is, not the fittest will survive, contrary to what Darwin's evolution theory asserts. He asks: If life's tape is replayed, will history repeat itself and humans can still be found on earth? His own answer is "no." Debates go on but nothing is done seriously and scientifically. Worse yet, there is no second data point forthcoming. Our active-walk aggregation (AWA) model [Lam & Pochy, 1993; Lam, 2005a] is able to shed light on this debate. The AWA model says "maybe" in answering Gould's question [Lam, 1998]. It is "maybe" because if the world lies in the "sensitive zone",[6] then the growth outcome may not be repeatable; otherwise, it is repeatable, more or less. The problem is to know where our world sits.

3. Modeling social history: will all societies end up as liberal democratic societies?

Francis Fukuyama, considered one of fifty key thinkers on history, publishes in 1989 an article, "End of History?" [Fukuyama, 1989]. He asserts that every human being needs two satisfactions, namely, economic well being and "recognition," with the latter meaning respect by others. He argues that since the liberal democratic society is the only one that can satisfy its citizens on these two basic needs, consequently, given enough time, all societies will end up as liberal democratic societies. And that will be the end of history, if history is understood to be the directional change in societal forms. Misunderstandings of Fukuyama's thesis ensure and debates go on in the history profession. Nothing is done scientifically to settle the issue.

---

[6] The sensitive zone is a region in the parameter space, within which, for the same set of parameters, different runs of the computer model may result in different patterns due to the use of a different sequence of random numbers in each run.

In our view, the two human needs suggested by Fukuyama should be generalized to "body satisfaction" and "soul satisfaction." After all, body and "soul" (or spirit) comprise the whole of a human being. And we know for sure, for example, when someone joins a revolution to change the society, the person may give up her or his life before the revolution succeeds, if at all—and recognition is not in the person's mind. The degrees of satisfaction of "body" and "soul" in each society can be quantified by two indices, obtained from a survey of its citizens. To test Fukuyama's thesis, one can represent each society as an active walker, a particle, moving in the two-dimensional space of "body" and "soul" indices (see Fig. 13.3) [Lam, 2002]. At each point in this space, a "fitness" potential can be defined. The movement of each particle (usually, but not always, up the scales) will change the fitness landscape and influence the movement of other particles. The problem will be to find out, under what circumstances, all the particles will cluster together at the location corresponding to a liberal democratic society. It is thus a problem of clustering of active walkers in a two-dimensional deformable landscape.[7]

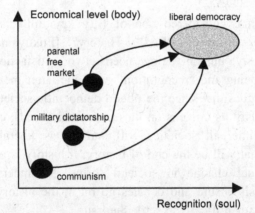

Fig. 13.3. Sketch of an active walk model for the evolution of political systems.

---

[7] The model can be generalized to include the possibility that two particles may combine into one, and some particles may split into two or three—some kind of chemical reactions—corresponding to the case in history that countries may get unified or fragmented in time.

Such a problem has been studied before in physics in another context, and clustering of active walkers indeed occurs [Schweitzer & Schimansky-Geier, 1994]. The corresponding investigation in history as outlined above will bring Fukuyama's historical study one level up, as scientific level is concerned, and serve as an example in other cases.

Note that AW is not the only paradigm possible in modeling history. And, in rare occasions, modeling of a system can be done analytically without the use of computers.

### 13.2.3  *Computer Simulation*

Another approach is the method of computer simulations, which usually are at the realistic level—with the mechanisms incorporated, even only simplified mechanisms. Here is a very interesting example.

Simulating the growth of a historical society

A simulation of the development of a society in the Long House Valley in the Black Mesa area of northeastern Arizona, USA, was carried out by Axtell *et al.* [2002]. The simulation results show agreement with the quantitative historical data, which are reconstructed from paleoenvironmental research based on alluvial geomorphology, palynology and dendroclimatology. For example, between the years *anno Domini* 400-1400, the number of households has two peaks; this is reproduced in the simulation. So is the evolution of the spatial distribution of settlement. In this study, heterogeneity in both agents and the landscape, hard to model mathematically, is found to be crucial. The modeling starts with a landscape reconstructed from paleoenvironmental variables, which is then populated with artificial agents representing individual households. Five household attributes are specified, together with household rules guessed from historical data. The model involves 14 reasonably chosen parameters, plus eight adjustable parameters for optimization. The model is very detailed. It is interesting to see whether the model can be simplified to its bone, with fewer parameters, that can still produce essentially the same results.

### 13.2.4 *The Zipf Plot*

Sometimes, very interesting results can be obtained from some very simple techniques which are quite well known in the field of complex systems. An example is the use of *Zipf plot*, which is at the empirical level. To obtain this plot for a given sequence of numbers, there are four steps:

1. The sequence of numbers is rearranged in a decreasing order.
2. Redundancy is removed by keeping only one number among those of same magnitude, resulting in a sequence of monotonic decreasing numbers.[8]
3. The largest number is assigned rank 1, the second largest rank 2, etc.
4. The Zipf plot is the curve appearing in the plot of the number *vs.* rank (with rank as the horizontal axis).

Here is an example concerning the Chinese dynasty lifetimes [Lam, 2006a; 2006b]. The lifetime of a Chinese dynasty $\tau_D$ is the sum of regime lifetimes $\tau_R$, corresponding to all the emperors within the same dynasty. The Zipf plot of $\tau_D$ is given in Fig. 13.4, with the presence of 26 data points, less than 31 (the number of available dynasties), due to the adopted procedure of removing redundancies in the data sequence. The data points fall on two straight lines—a result named the *bilinear effect* [Lam, 2006b; Lam *et al.*, 2008]. It implies that

1. The "curse of history," as Chinese dynasties are concerned, does exist.
2. A dynasty can survive every $3.5 \pm 0.1$ years if it lasts $57 \pm 2$ years or less; beyond that, every $25.6 \pm 0.1$ years—dynasty lifetime is discrete, or "quantized."
3. There is a transition point separating these two different behaviors.

---

[8] Other people might retain all the numbers of the same magnitude in the sequence, resulting in a Zipf plot which could contain horizontal parts. Our version here is more reasonable, because one would like to fit the plot to a smooth curve.

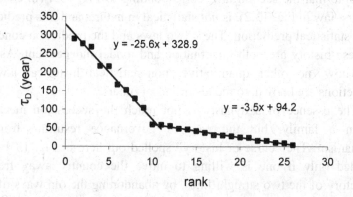

Fig. 13.4. The Zipf plot of Chinese dynasty lifetime $\tau_D$, an example of the bilinear effect.

This is *the second quantitative law concerning Chinese history*. Whether the discreteness in $\tau_D$ results from some periodic external conditions in the Chinese history or is a self-organizing phenomenon resulting from some nonlinear dynamics remains to be investigated.

The bilinear effect is the phenomenon that an adaptive system becomes stronger after existing for a period of time. The mere fact of survival reinforces its strength, through learning, restructuring, and so on. Similar behavior is known to exist in the case of restaurants, corporations or biological species. What is surprising here is the presence of two linear lines and a sharp transition point (in the Zipf plot).

A *quantitative* prediction could be inferred from Fig. 13.4. Under the *assumption* that Chinese dynasties remain in the bilinear-effect class, any dynasty after Qing, if exists, will either

1. last $303 \pm 1$ years or less, and fall more or less on the two lines in Fig. 13.4; or
2. end definitely and exactly in its year $329 \pm 1$.[9]

---

[9] The number 303 is the height at rank 1 on the straight line in Fig. 13.4; 329, that at "rank" 0.

Note that the second law, corresponding to Fig. 13.4 (in contrast to the first law in Fig. 13.2), is *not* statistical in nature; and this prediction is not a statistical prediction. These two laws and the prediction concerning Chinese history are both quantitative and model independent. As far as we know, no other quantitative, non-statistical historical laws and predictions are known, to the historians or others.

The essence of a dynasty is not much the succession mechanism within a family, but the way of governance resulting from that mechanism. The "curse of history" spelled out here in Fig. 13.4 can be avoided only if one is willing to move the country away from the trajectory of the two straight lines, by abandoning the old ways of doing things.

It turns out that this regularity in Chinese history is only a particular case of the bilinear effect; more examples are subsequently found in other human affairs and complex systems (Fig. 13.5) [Lam *et al.*, 2008]. Figure 13.5a is the Zipf plot of the number of votes for Chinese *xiaopin* actors.[10] Figure 13.5b comes from the airline quality ratings in the year 2005.[11]

In other words, the bilinear effect is a *new* class of Zipf plots, apart from the other two well-known classes—power laws [Newman, 2005] and stretched-exponent distributions [Laherrère & Sornette, 1998].

The generic mechanism behind bilinear effect is not yet understood. But we do know more than one way to obtain the bilinear effect [Lam *et al.*, 2008]. For example, first pick $N_p$ number of points according to a one-peak probability distribution function $p(x)$, say, to obtain a sequence of numbers $\{x_i\}$, with $i = 1, 2, ..., N_p$. That is, the probability that $x_i$ being picked is proportional to $p(x_i)$. A Zipf plot is then performed with this sequence $\{x_i\}$. With luck, the Zipf plot is bilinear.[12] The chance of obtaining bilinear effect increases as $N_p$ is increased. Such a mechanism is applicable to the case in Fig. 13.5a, wherein, the votes were cast

---

[10] ent.sin.com.cn/2004-09-30/1050521359.html (Oct. 7, 2004). *Xiaopin* is a popular form of short drama performed by a cast of usually two actors in China.

[11] www.aqr.aero.

[12] Experience shows that the one-peak shape of $p(x)$ is not a necessity in obtaining bilinear effect this way. Sometimes, a monotonic decreasing $p(x)$ also works; but so far, a decreasing power-law $p(x)$ does not seem to work.

*independently* by people on the Web from a *xiaopin* list. It is not applicable to the Chinese dynasties, the lifetimes of which are obviously correlated.

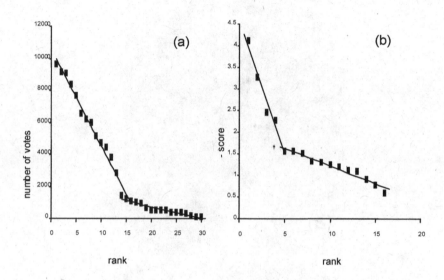

Fig. 13.5. Two additional examples of the bilinear effect. Zipf plots of (a) popular votes for *xiaopin* actors, and (b) airline quality data.

## 13.3 History in the Future

The importance of history can be seen, for example, through its negative impact on human lives. Powerful political leaders could mistake an unproven historical hypothesis as firm theory, apply it to a confined population and cause millions of death in a few short years [Lam, 2002]. Another example is the massive protests in China few years ago, due to different interpretation of past history involving two countries (Fig. 13.6). Yet, in spite of its importance, the physical basis of history is unrecognized by most historians. For instance, in the historiography textbook *The New Nature of History* [Marwick, 2001] the alleged "fundamental differences" between history and the sciences are listed:

Fig. 13.6. Protests in China, April 2005. (a) "FACE HISTORY" is the slogan on the left placard. (b) "PROTECT DIAOYUDAO" is a historical issue also raised in the protests. Diaoyudao, or diaoyutai, is a group of tiny islands in the East China Sea. The "protect diaoyutai" movement was started by oversea Chinese students in the United States at the end of 1970 [The Seventies Monthly, 1971].

1. Fundamental difference in the subject of study: natural sciences concern natural world and physical world; history concerns human beings and human society, very different in character.
2. No controlled experiments by historians.
3. Historians develop theories and theses, but not concerned with developing laws and theories like that in sciences.
4. History studies do not have prediction power.
5. Relations and interactions in history studies are not expressed mathematically.
6. Historians report their findings in prose (articles or books), not in terse research articles.

Unfortunately, all six points are wrong, for the following reasons.

1. As explained in Section 13.1, human beings and thus human society are material systems, which are part of the natural sciences. Human society share same characteristics as other inanimate complex systems, as demonstrated in Figs. 13.1, 13.2, 13.4 and 13.5.
2. Some physical disciplines like astronomy and archeology also do not have controlled experiments.
3. It is untrue that all historians are not concerned with developing laws and theories in history. Some tried, not very successfully, partly due to their inadequate training in using scientific tools. Historical laws do exist, as shown in Sections 13.2.1 and 13.2.4 and in Figs. 13.1, 13.2 and 13.4.
4. History studies, like that in Section 13.2.4, do have prediction power.
5. Relations and interactions in history studies can be expressed mathematically. An example is the landscape theory of Axelrod and Bennett [1993] to show how and why 17 European nations in World War II aligned themselves into two large groups. The pairwise propensities between nations are assigned numerical values, and the configuration energy in the (fixed) landscape is given in equations.[13]

---

[13] See [Galam, 1998] for a comment on this work, and the following response by the original authors.

6. Historians do report their findings in research articles, terse or not. That is why there exist quite a number of history journals, such as *History and Theory* and *American Historical Review*. It is true that many historians still skip the peer-reviewed journals and directly report their findings in books—not a healthy thing for the history discipline, epistemologically speaking [Lam, 2002]. These are actually popular history books, like the popular science books written by physicists.[14] In the case of the history profession, there are at least three reasons behind this practice. (1) Many research results in history are still at the data gathering and empirical analysis stage, not very technical and can be presented in narratives. (2) There is enough number of readers out there who is willing to pay to find out what happened to their ancestors or their own kind in the past. In contrast, not that many will pay to learn what happened to the electrons. Bad for physics. (3) Historically, before history became a professional discipline in the universities in the second half of the nineteenth century, historians had to earn their living by writing books that are readable and salable to the public [Stanford, 1998]. In other words, writing popular history books was a survival need for historians, a tradition carried over up to now.

These errors are due to misunderstanding of the nature of science, and the neglect of the material basis of the historical system itself. The inadequate science training received by historians, past and present, explains why they failed to find historical laws. For example, the Chinese dynasty data have been lying there after 1912; the plots in Figs. 13.2 and 13.4 could be carried out by hand without computers, and even by high school students. But unless one knows about power laws and the existence of the Zipf plot, there is no motivation to do so. And these are current topics in the study of complex systems. Ironically, Zipf plots were first done by George Zipf (1902-1950), a Harvard linguist, with data from the humanities and social sciences.

---

[14] The unique characteristics of popular-science books and how to integrate them into science teaching are discussed in [Lam, 2001; 2005b; 2006c; 2008b].

Quite recently and surprisingly, while the importance of history is well recognized in Hong Kong, the history department of the University of Hong Kong is threatened with closure because it fails to attract enough number of students [Xie, 2005]. Anyway, it is time for all history departments to revamp their curriculum, by increasing the mathematical skills of their students, going beyond story telling and making history research more technical and scientific, and creating a course on the physics of history (or *histophysics* [Lam, 2002; 2004]). This revamp will help current students to become better historians after they graduate, and may appeal to a new class of incoming students who have a technical background but feel attracted more to the humanities than the traditional sciences.

## 13.4 Conclusion

As shown above, human history can indeed be studied scientifically, using techniques borrowed from physics and complex systems. Human history is thus a true example of Science Matters (SciMat), the new discipline that treats all human-related matters as part of science [Lam, 2008a; 2008c]. This, of course, does not imply that the conventional studies by other historians or the existing history departments should be abolished. On the contrary, they are very valuable. They are doing splendid jobs at the *empirical level*, the first level in the scientific development of any discipline, which is to collect data, analyze and summarize data, and come up with "explanations" in understanding them. Yet, the explanations are usually educated guesses—"hypotheses" but not yet "theories," in the sense that a theory is a confirmed hypothesis. The single-event, unrepeatable nature of history makes the direct confirmation of a hypothesis very difficult, unlike the case in many "natural sciences." It is at this point that history as a SciMat enters, as illustrated in Section 13.2.

As demonstrated, with a little bit of luck and the right perspective and right tool, a hidden historical *law* (not merely a historical trend) might suddenly jump out and meet the eyes of the investigator. With a lot of luck, this historical law might even lead one to the discovery of a general principle in Nature (such as the bilinear effect).[15]

What we try to do is to raise historical studies to a higher scientific level—to the phenomenological and realistic levels. To this end, collaborations between traditional historians and physicists are strongly urged. And everybody gains.

# References

Arthur, B. [1990] "Positive feedbacks in economy," *Sci. Am.*, Feb., 92.

Axelrod, R. & Bennett, D. S. [1993] "A landscape theory of aggregation," *British J. Political Sci.* **23**, 211-233.

Axtell, R. L., Epstein, J. M, Dean J. S., Gumerman, G. J., Swedlund, A. C., Harburger, J., Chakravarty, S., Hammond, R., Parker, J. & Parker, M. [2002] "Population growth and collapse in a multiagent model of the Kayenta Anasazi in Long House Valley," *Proc. Natl. Acad. Sci. USA* **99**, Suppl. 3, 7275-7279.

David, P. A. [1986] "Understanding the economics of QWERTY: The necessity of history," in *Economic History and the Modern Economist*, ed. Parker, W.N. (Blackwell, New York) pp. 30-49.

Feder, T. [2005] "Lab webs brain research and physics," *Phys. Today*, April, 26-27.

Fukuyama, F. [1989] "The end of history?" *The National Interest* **16** (summer), 3-18.

Galam, S. [1998] "Comment on 'A landscape theory of aggregation'," *British J. Political Sci.* **28**, 411-412.

Gardiner, P. (ed.) [1959] *Theories of History* (The Free Press, Glencoe, IL).

Gould, S. J. [1989] *Wonderful Life: The Burgess Shale and the Nature of History* (Norton, New York).

Han, X.-P., Hu, C.-D., Liu, Z.-M. & Wang, B.-H. [2008] "Parameter-tuning networks: Experiments and active-walk model," *Euro. Phys. Lett.* **83**, 28003.

Huang, R. [1997] *China: A Macro History* (M.E. Sharpe, Armonk, NY).

Johnston, A. C. & Nava, S. [1985] "Recurrence rates and probability estimates for the New Madrid seismic zone," *J. Geophys. Res.* **90**, 6737-6753.

---

[15] Starting with a particular case and ending with a general principle is the common route of discoveries in physics and other disciplines. We ourselves have experienced this before: the modeling of filamentary patterns found in thin cells of electrodeposit experiments led us to active walks, a general paradigm for complex systems [Lam, 2005a; 2006b]. Similarly, that was how Charles Darwin (1809-1882) found his evolutionary principle for living systems.

Krakauer, D. C. [2007] "The quest for patterns in meta-history," *Santa Fe Inst. Bull.,* Winter, 32-39.

Laherrère, J. & Sornette, D. [1998] "Stretched exponential distributions in nature and economy: 'fat tails' with characteristic scales," *Eur. Phys. J. B.* **2**, 525-539.

Lam, L. & Pochy, R. D. [1993] "Active walker models: Growth and form in nonequilibrium systems," *Comput. Phys.* **7**, 534-541.

Lam, L. [1998] *Nonlinear Physics for Beginners: Fractals, Chaos, Solitons, Pattern Formation, Cellular Automata and Complex Systems* (World Scientific, Singapore).

Lam, L. [2001] "Raising the scientific literacy of the population: A simple tactic and a global strategy," in *Public Understanding of Science,* ed. Editorial Committee (Science and Technology University of China Press, Hefei, China).

Lam, L. [2002] "Histophysics: A new discipline," *Mod. Phys. Lett. B* **16**, 1163-1176.

Lam, L. [2004] *This Pale Blue Dot: Science, History, God* (Tamkang University Press, Tamsui).

Lam, L. [2005a] "Active walks: The first twelve years (Part I)," *Int. J. Bifurcation and Chaos* **15**, 2317-2348.

Lam, L. [2005b] "Integrating popular science books into college science teaching," *The Pantaneto Forum,* Issue 19.

Lam, L. [2006a] "How long can a Chinese dynasty last?" (preprint).

Lam, L. [2006b] "Active walks: The first twelve years (Part II)," *Int. J. Bifurcation and Chaos* **16**, 239-268.

Lam, L. [2006c] "Science communication: What every scientist can do and a physicist's experience," *Science Popularization,* No. 2, 36-41. See also Lam, L., in *Proceedings of Beijing PCST Working Symposium,* June 22-23, 2005, Beijing, China.

Lam, L. [2008a] "Science Matters: A unified perspective," in *Science Matters: Humanities as Complex Systems,* eds. Burguete, M. & Lam, L. (World Scientific, Singapore).

Lam, L. [2008b] "SciComm, PopSci and The Real World," in *Science Matters: Humanities as Complex Systems,* eds. Burguete, M. & Lam, L. (World Scientific, Singapore).

Lam, L. [2008c] "Science Matters: The newest and biggest interdicipline," in *China Interdisciplinary Science,* Vol. 2, ed. Liu, Z.-L. (Science Press, Beijing).

Lam, L., Bellavia, D. C., Han, X. P., Liu, A., Shu, C. Q., Wei, Z. J., Zhu, J. C. & Zhou, T. [2008] "Bilinear effect in complex systems" (preprint).

Levy, J. S. [1983] *War in the Modern Great Power System, 1495-1975* (University Press of Kentucky, Lexington).

Marvick, A. [2001] *The New Nature of History* (Lyceum, Chicago) p. 248.

Morby, J. E. [2002] *Dynasties of the World* (Oxford University Press, Oxford).

Newman, M. E. J. [2005] "Power laws, Pareto distributions and Zipf's law," *Comtemp. Phys.* **46**, 323-351.

Paul, W. & Baschnagel, J. [1999] *Stochastic Processes: From Physics to Finance* (Springer, New York).

Richardson, L. F. [1941] "Frequency of occurrence of wars and other fatal quarrels," *Nature* **148**, 598.

Roberts, D. C. & Turcotte, D. L. [1998] "Fractality and self-organized criticality of wars,' *Fractals* **6**, 351-357.

Schweitzer, F. & Schimansky-Geier, L. [1994] "Clustering of active walkers in a two-component system," *Physica A* **206**, 359-379.

Sornette, D. [2000] *Critical Phenomena in Natural Sciences* (Springer, New York).

Stanford, M. [1998] *An Introduction to the Philosophy of History* (Blackwell, Malden, MA) p. 228.

The Seventies Monthly (ed.) [1971] *Truth Behind the Diaoyutai Incident* (The Seventies Monthly, Hong Kong).

Xie, Xi [2005] "Historical feel regained," *World Journal* (Millbrae, CA), April 16, A15.

Zipf, G.K. [1949] *Human Behavior and the Principle of Least Effort* (Addison-Wesley, Cambridge, MA).

# Contributors

**Maria Burguete** received her Ph.D. in History of Science (contemporary chemistry) from Ludwig Maximilians University at Munich, Germany (2000). She was the very first biochemist to graduate from the Faculty of Sciences in Lisbon (1982), after completing a Bachelor Degree in Chemical Engineering (1979) at the Lisbon Higher Institute of Engineering (ISEL). She is a scientist with teaching and research experience in a wide variety of scientific fields. This diversity enhanced the development of both her interdisciplinarity and a transdisciplinarity. She is now a scientist at Bento da Rocha Cabral in Portugal. She has published five scientific books and five poetry books, and over 20 scientific papers mostly in history and philosophy of science. *Email: mariabuguete@ gmail.com*

**Paul Caro** is a former (retired) Director of Research at CNRS who has worked for many years in inorganic chemistry. He is a rare-earths specialist. In the 1980s he became interested in science popularization through newspaper articles (in "Le Monde" and magazines), radio broadcasts (France Culture, Radio Classique, mostly), television shows (TF1), exhibitions in Museums and some books. He was until 2001 in charge of "scientific affairs" at the Cité des Sciences et de l'Industrie in Paris. He has served as an adviser for several DG Research European programs in the field of education and is also a scientific adviser for the Portuguese Program "Ciência Viva." He is a Corresponding Member of the French Academy of Sciences and a Member of the French Academy of Technology. *Email: paul_caro@hotmail.com.*

**Alfredo Dinis** graduated in Philosophy at the Catholic University of Portugal (1979) and in Theology at the Gregorian University of Rome (1985). He obtained an M.Phil. and a Ph.D. in History and Philosophy of Science from the University of Cambridge, UK (1986 and 1989). Since 1989 he has been lecturing in Philosophy of Science, Logic, and Cognitive Science at the Faculty of Philosophy of Braga (Portugal), where he is also at present the Faculty Dean. He is the President of the Portuguese Society for the Cognitive Sciences. His present research is centered upon the philosophical, ethical and theological consequences of recent developments in biology and in the cognitive sciences. *Email: adinis@braga.ucp.pt.*

**Xiao-Pu Han**, master degree graduate student at the University of Science and Technology of China, obtained his B.Sc. from Shandong University. He has published 15 scientific papers in research journals and conference proceedings since 2004. His current research interest focuses on human dynamics and complex networks. *Email: hxp@mail.ustc. edu.cn.*

**Brigitte Hoppe** obtained her state diploma in pharmaceutical sciences at the University of Freiburg (Breisgau); she earned the degree of Dr. phil. nat. in the History of Natural Sciences at the University of Frankfurt (Main) in 1964. Her research on epistemological changes in biology in Early Modern Times was the basis of the habilitation thesis at the University of Munich in 1972. As an Associate Professor at this university (1980), she was responsible for a working group in the History of Life Sciences. Dr. Hoppe published 7 books and more than 200 papers. She is an Effective Member of the International Academy for History of Sciences and member of national and international societies in this field. Her current research is in the history of biological sciences from 17th to 20th centuries, and on explorations in the field of natural history in overseas countries. *Email: B.Hoppe@lrz.uni-muenchen.de.*

**Lui Lam** obtained his B.Sc. (with First Class Honors) from the University of Hong Kong, M.Sc. from University of British Columbia, and Ph.D. from Columbia University. He is Professor of Physics at San Jose State University, California and Adjunct Professor at both the Chinese Academy of Sciences and the China Association for Science and Technology. Prof. Lam invented bowlics (1982), one of three existing types of liquid crystals in the world; active walks (1992), a new paradigm in complex systems; and a new discipline called histophysics (2002). Lam published 11 books and over 160 scientific papers. He is the founder of the International Liquid Crystal Society (1990); cofounder of the Chinese Liquid Crystal Society (1980); founder and editor-in-chief of the Springer book series *Partially Ordered Systems*. His current research is in histophysics, complex systems and science matters. *Email: lui2002lam@yahoo.com.*

**Da-Guang Li**, a professor at the Graduate University of Chinese Academy of Sciences, is the Executive Director of the 1996, 2001 and 2003 Surveys of Public Understanding of Science of Chinese Adults, and the Chinese translator of the popular science book *The Demon Haunted World* by Carl Sagan. *Email: ldaguang@yahoo.com.*

**Bing Liu** obtained his B.Sc. from Peking University and M.Sc. from the Chinese Academy of Sciences. He is now Professor of History and Philosophy of Science, in the STS Institute, School of Humanities and Social Sciences, Tsinghua University, Beijing, China. Prof. Liu has published more than 20 books and over 170 papers in history and philosophy. His current research is in history of science, and science communication. *Email: liubing@tsinghua.edu.cn.*

**Dun Liu**, the former Director of the Institute for the History of Natural Science within the Chinese Academy of Sciences (1997-2005), is the incumbent President of the Chinese Society for the History of Science and Technology. Among his numerous publications, the book *How Great Math Is* (Shenyang, 1993) and several treatises on Chinese mathematics/astronomy within the social context of the Ming to Qing period (ca. $16^{th}$-$17^{th}$ centuries) are most appreciated by researchers. He is editor-in-chief of the bimonthly journal, *Science & Culture Review*. His current focus is on historical and cultural issues of science. *Email: liudun@ustc.edu.cn.*

**Nigel Sanitt** obtained his B.Sc. in Physics from Imperial College, London and Part III of the Mathematics Tripos and Ph.D. from Cambridge University, where he trained as an astrophysicist at the Institute of Astronomy. He is founder and editor of *The Pantaneto Forum*, a journal which aims to promote debate on how scientists communicate, with particular emphasis on how such communication and research skills can be improved through a better philosophical understanding of science. His book *Science as a Questioning Process* was published in 1996, and he has edited a collection of articles from the first five years of The Pantaneto Forum under the title: *Motivating Science. Email: nigel@ pantaneto.co.uk.*

**Michael Shermer** is Publisher of *Skeptic* magazine (www.skeptic.com) and a monthly columnist for *Scientific American* (www.sciam.com). He is the author of *Why People Believe Weird Things, How We Believe, The Science of Good and Evil*, and *Why Darwin Matters*. His new book is *The Mind of the Market: Compassionate Apes, Competitive Humans, and Other Tales from Evolutionary Economics* (Henry Holt/Times Books). *Email: mshermer@skeptic.com.*

**Bing-Hong Wang**, graduated from Theoretical Physics Specialty, Department of Modern Physics, University of Science and Technology of China (USTC), Beijing, in 1967, did his Post Doctoral Research in Department of Physics, Stevens Institute of Technology, USA during 1982-1985. His current occupation: Professor, Ph.D. Doctor's Adviser of Theoretical Physics, USTC; Director, Institute of Theoretical Physics, USTC; Head, National Key Important Specialty of Theoretical Physics, USTC; Director, Institute of Complex Adaptive System, Shanghai Academy of System Science, Shanghai, China; Executive Director, Nonlinear Science Center, USTC; and President, Nonlinear Science Society of Anhui Province, P. R. China. His recent research interests focus on traffic flow, statistical physics, nonlinear dynamics, econophysics, complex networks and complex adaptive system theory. He has published more than 200 research papers in scientific journals. *Email: bhwang@ ustc.edu.cn.*

**Tao Zhou** obtained his B.Sc. from the University of Science and Technology of China (USTC) in 2005, and is a Ph.D. student majoring in theoretical physics at USTC now. He has published over 100 papers in refereed journals, with more than 70 of them included in *Web of Science* (SCI Index). His publications together obtained more than 500 citations according to *Web of Science*. His current research interests include complex networks, human dynamics, econophysics, statistical mechanics of information systems, and collective dynamics of self-driven particles. *Email: zhutou@ustc.edu.*

# Index